Everyday Mathematics®

The University of Chicago School Mathematics Project

Home Links

Grade **3**

 **Wright Group**

The **McGraw·Hill** Companies

The University of Chicago School Mathematics Project (UCSMP)

Max Bell, Director, UCSMP Elementary Materials Component; Director, *Everyday Mathematics* First Edition; James McBride, Director, *Everyday Mathematics* Second Edition; Andy Isaacs, Director, *Everyday Mathematics* Third Edition; Amy Dillard, Associate Director, *Everyday Mathematics* Third Edition

Authors

Max Bell, Jean Bell, John Bretzlauf, Mary Ellen Dairyko*, Amy Dillard, Robert Hartfield, Andy Isaacs, James McBride, Kathleen Pitvorec, Peter Saecker

Third Edition only

Technical Art

Diana Barrie

Teachers in Residence

Lisa Bernstein, Carole Skalinder

Editorial Assistant

Jamie Montague Callister

Contributors

Carol Arkin, Robert Balfanz, Sharlean Brooks, James Flanders, David Garcia, Rita Gronbach, Deborah Arron Leslie, Curtis Lieneck, Diana Marino, Mary Moley, William D. Pattison, William Salvato, Jean Marie Sweigart, Leeann Will

Photo Credits

©C Squared Studios/Getty Images, p. 156; ©Tim Flach/Getty Images, cover; Getty Images, cover, *bottom left*; ©JupiterImages/Comstock, p. 76; ©Ken O'Donoghue, pp. 177, 178; Royalty-free/Corbis, pp. 227, 252.

www.WrightGroup.com

Printed in the United States of America.

Send all inquiries to:
Wright Group/McGraw-Hill
P.O. Box 812960
Chicago, IL 60681

ISBN-13 978-0-07-609740-1
ISBN-10 0-07-609740-4

2 3 4 5 6 7 8 9 DBH 12 11 10 09 08 07

Contents

Contents **v**

HOME LINK 1·1 | **Unit 1: Family Letter**

Introduction to Third Grade Everyday Mathematics®

Welcome to *Third Grade Everyday Mathematics.* It is part of an elementary school mathematics curriculum developed by the University of Chicago School Mathematics Project. *Everyday Mathematics* offers children a broad background in mathematics.

Several features of the program are described below to help familiarize you with the structure and expectations of *Everyday Mathematics.*

A problem-solving approach based on everyday situations By making connections between their own knowledge and their experiences, both in school and outside of school, children learn basic math skills in meaningful contexts so that the mathematics becomes real.

Frequent practice of basic skills Instead of practice presented in a single, tedious drill format, children practice basic skills in more engaging ways. In addition to completing daily review exercises covering a variety of topics, children work with patterns on a number grid, and solve addition and subtraction fact families in different formats. Children will also play games that are specifically designed to develop basic skills.

An instructional approach that revisits concepts regularly To enhance the development of basic skills and concepts, children regularly revisit concepts and repeatedly practice skills encountered earlier. The lessons are designed to build on previously learned concepts and skills throughout the year instead of treating them as isolated bits of knowledge.

A curriculum that explores mathematical content beyond basic arithmetic Mathematics standards around the world indicate that basic arithmetic skills are only the beginning of the mathematical knowledge children will need as they develop critical thinking skills. In addition to basic arithmetic, *Everyday Mathematics* develops concepts and skills in the following topics—number and numeration; operations and computation; data and chance; geometry; measurement and reference frames; and patterns, functions, and algebra.

1

Third Grade Everyday Mathematics emphasizes the following content:

Number and Numeration Counting patterns; place value; reading and writing whole numbers through 1,000,000; fractions, decimals, and integers

Operations and Computation Multiplication and division facts extended to multidigit problems; working with properties; operations with fractions and money

Data and Chance Collecting, organizing, and displaying data using tables, charts, and graphs; using basic probability terms

Geometry Exploring 2- and 3-dimensional shapes and other geometric concepts

Measurement Recording equivalent units of length; recognizing appropriate units of measure; finding the areas of rectangles by counting squares

Reference Frames Using multiplication arrays, coordinate grids, thermometers, clocks, calendars; and map scales to estimate distances

Patterns, Functions, and Algebra Finding patterns on the number grid; solving Frames-and-Arrows puzzles having two rules; completing variations of "What's My Rule?" activities; exploring the relationship between multiplication and division; using parentheses in writing number models; naming missing parts of number models

Everyday Mathematics will provide you with ample opportunities to monitor your child's progress and to participate in your child's mathematics experiences.

Throughout the year, you will receive Family Letters to keep you informed of the mathematical content your child will be studying in each unit. Each letter will include a vocabulary list, suggested Do-Anytime Activities for you and your child, and an answer guide to selected Home Link (homework) activities.

You will enjoy seeing your child's confidence and comprehension soar as he or she connects mathematics to everyday life. We look forward to an exciting year!

Routines, Review, and Assessment

The first purpose of Unit 1 is to establish routines that children will use throughout the school year. The second purpose is to review and extend mathematical concepts that were developed in previous grades.

In Unit 1, children will look for examples of numbers for the Numbers All Around Museum. Examples of numbers might include identification numbers, measures, money, telephone numbers, addresses, and codes. Children will also look at number patterns in a problem-solving setting by using number-grid puzzles and Frames-and-Arrows diagrams. (*See examples on the next page.*)

Throughout Unit 1, children will use numbers within the context of real-life situations. After reviewing place-value concepts, children will work with money and pretend to purchase items from a vending machine and a store. The emphasis on applying numbers to the real world is also reflected in the yearlong Length-of-Day Project, a weekly routine that involves collecting, recording, and graphing sunrise/sunset data.

Vocabulary

Important terms in Unit 1:

digits Any of the symbols 0, 1, 2, 3, 4, 5, 6, 7, 8, and 9 in the base 10 numeration system.

estimate An answer close to, or approximating, an exact answer.

tool kits In *Everyday Mathematics,* a bag or box containing a calculator, measuring tools, and manipulatives often used by students of the program.

number grid In *Everyday Mathematics,* a table in which consecutive numbers are arranged, usually in 10 columns per row. A move from one number to the next within a row is a change of 1; a move from one number to the next within a column is a change of 10.

									0
1	2	3	4	5	6	7	8	9	10
11	12	13	14	15	16	17	18	19	20
21	22	23	24	25	26	27	28	29	30

number-grid puzzle In *Everyday Mathematics,* a piece of the number grid in which some, but not all, of the numbers are missing. Children use number-grid puzzles to practice place-value concepts.

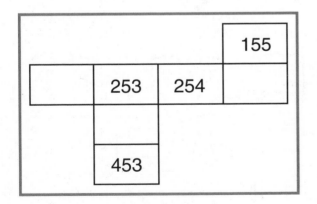

range The difference between the *maximum* and the *minimum* in a set of data. Used as a measure of the spread of data.

mode The value or values that occur most often in a set of data.

name-collection box In *Everyday Mathematics,* a diagram that is used for collecting equivalent names for a number.

300

three hundred	310 − 10
150 + 150	260 + 40
trescientos	300 − 0

Frames-and-Arrows In *Everyday Mathematics,* diagrams consisting of frames connected by arrows used to represent number sequences. Each frame contains a number and each arrow represents a rule that determines which number goes in the next frame. There may be more than one rule, represented by different colored arrows.

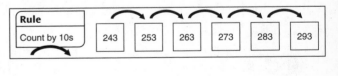

3

As You Help Your Child with Homework

As your child brings home assignments, you may want to go over the instructions together, clarifying them as necessary. The answers listed below will guide you through this unit's Home Links.

Home Link 1·1

1. Answers vary.   **2.** 7; 7; 7; 7

Home Link 1·2

1. 21; 41   **2.** 164; 166; 184; 186
3. Sample answers: 97; 98; 99; 100; 108; 119; 127; 128; 129; 130
4. 1,372; 1,383; 1,392; 1,393; 1,394

Home Link 1·3

Sample answers:

1. ②, 4 ⑥ 7   **2.** 2,567   **3.** 2,367   **4.** 899; 908; 910
5. 1,044; 1,055; 1,065   **6.** 9   **7.** 4   **8.** 9   **9.** 5

Home Link 1·4

1. Answers vary.   **2.** 8:00   **3.** 3:30   **4.** 6:15
5. 11:45   **6.** 7:10   **7.** 5:40   **8.** Answers vary.

Home Link 1·5

1.

Time Spent Watching TV	
Hours	Children
0	/
1	//
2	//
3	////
4	/
5	/

2. 0   **3.** 5   **4.** 5   **5.** 3   **6.** 3

Home Link 1·6

1.

18	Sample answers:
9 + 9	2 × 9
6 + 6 + 6	~~HHT HHT HHT~~ ///
dieciocho 4 × 5 − 2	36 ÷ 2
number of days in two weeks + 4 days	

2.

12	~~HHT HHT~~	one dozen
	7 + 5	
	number of months in 1 year	
	15 − 3	10 + 2
	~~13~~	~~7~~

3. Answers vary.

Home Link 1·7

Sample answers:

1. sure to happen   **2.** sure not to happen
3. may happen, but not sure
4. may happen, but not sure   **5.** 7   **6.** 3
7. 4   **8.** 7

Home Link 1·8

1.

131	132	133	134	135	136	137	138	139	140
141	142	143	144	145	146	147	148	149	150
151	152	153	154	155	156	157	158	159	160
161	162	163	164	165	166	167	168	169	170
171	172	173	174	175	176	177	178	179	180

2. 154; 23   **3.** 148; 29   **4.** 22
5. Sample answer: I counted 2 tens from 180 and then 2 ones.   **6.** 6   **7.** 7   **8.** 13   **9.** 13

Home Link 1·9

1. , **2.** Answers vary.   **3.** 3   **4.** 3   **5.** 5   **6.** 3

Home Link 1·10

5. 6; 6; 5; 10   **6.** 6; 5; 2; 8

Home Link 1·11

3. 4   **4.** 11   **5.** 4   **6.** 11

Home Link 1·12

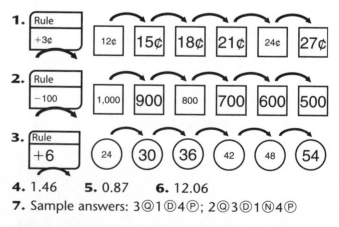

1. Rule +3¢: 12¢ 15¢ 18¢ 21¢ 24¢ 27¢
2. Rule −100: 1,000 900 800 700 600 500
3. Rule +6: 24 30 36 42 48 54

4. 1.46   **5.** 0.87   **6.** 12.06
7. Sample answers: 3 Ⓠ 1 Ⓓ 4 Ⓟ; 2 Ⓠ 3 Ⓓ 1 Ⓝ 4 Ⓟ

Home Link 1·13

4. 4   **5.** 4   **6.** 7   **7.** 7

HOME LINK 1·1 Numbers All Around Museum

Family Note

In the *Third Grade Everyday Mathematics* program, children *do* mathematics. We expect that children will want to share their enthusiasm for the mathematics activities they do in school with members of their families. Your child will bring home assignments and activities to do as homework throughout the year. These assignments, called Home Links, will be identified by the symbol at the top of the page. The assignments will not take very much time to complete, but most of them involve interaction with an adult or an older child.

There are good reasons for including Home Links in the third-grade program:

◆ The assignments encourage children to take initiative and responsibility for completing them. As you respond with encouragement and assistance, you help your child build independence and self-confidence.

◆ Home Links reinforce newly learned skills and concepts. They provide thinking and practice time at each child's own pace.

◆ These assignments are often designed to relate what is done in school to children's lives outside school. This helps tie mathematics to the real world, which is very important in the *Everyday Mathematics* program.

◆ The Home Links assignments will give you a better idea of the mathematics your child is learning in school.

Generally, you can help by listening and responding to your child's requests and comments about mathematics. You can help by linking numbers to real life, pointing out ways in which you use numbers (time, TV channels, page numbers, telephone numbers, bus routes, and so on). Extending the notion that "children who are read to, read," *Everyday Mathematics* supports the belief that children who have someone do math with them will learn mathematics. Playful counting and thinking games are very helpful in promoting such learning.

The Family Note will explain what the children are learning in class. Use it to help you understand where the assignment fits into your child's learning.

HOME LINK 1·1

Numbers All Around Museum *continued*

> **Family Note** Numbers on advertisements show quantities and prices (3 cans of soup for $1.00); food containers show weight or capacity (a $15\frac{1}{2}$-oz can of black beans or 1-quart carton of milk); and telephone books show addresses and phone numbers. By helping your child find examples of numbers in everyday life, you will reinforce the idea that numbers are all around us and are used for many reasons. Help your child recognize numbers by filling in the table.
>
> *Please return this Home Link to school within the next few days.*

1. Find as many different kinds of numbers as you can. Record the numbers in the table below. Be sure to include the unit if there is one.

Number	Unit (if there is one)	Where you found the number
14	oz	cereal box

Find objects or pictures with numbers on them to bring to school.
Check with an adult at home first. Do not bring anything valuable.

Practice

2. Solve.

$$\begin{array}{r} 5 \\ +2 \\ \hline \square \end{array} \qquad \begin{array}{r} \square \\ -2 \\ \hline 5 \end{array} \qquad \begin{array}{r} \square \\ -5 \\ \hline 2 \end{array} \qquad \begin{array}{r} 2 \\ +5 \\ \hline \square \end{array}$$

Number-Grid Puzzles

HOME LINK 1·2

Family Note   Today your child reviewed patterns on a number grid and completed number grid puzzles. On this Home Link, your child may use either the number grid or its patterns to complete the number grid puzzles. Ask your child to explain how he or she filled in the puzzles.

SRB 7–9

−9	−8	−7	−6	−5	−4	−3	−2	−1	0
1	2	3	4	5	6	7	8	9	10
11	12	13	14	15	16	17	18	19	20
21	22	23	24	25	26	27	28	29	30
31	32	33	34	35	36	37	38	39	40
41	42	43	44	45	46	47	48	49	50
51	52	53	54	55	56	57	58	59	60
61	62	63	64	65	66	67	68	69	70
71	72	73	74	75	76	77	78	79	80
81	82	83	84	85	86	87	88	89	90
91	92	93	94	95	96	97	98	99	100

When you move right, the numbers increase by 1.

When you move left, the numbers decrease by 1.

When you move down, the numbers increase by 10.

When you move up, the numbers decrease by 10.

Fill in the missing numbers. Explain the patterns to someone at home.

1.

	22
31	

2.

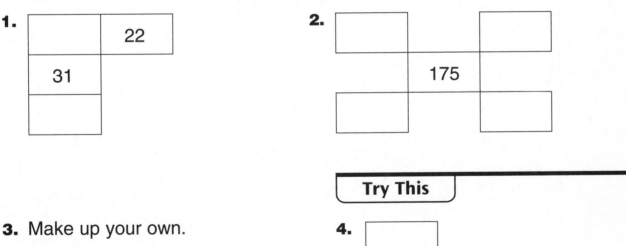

175

Try This

3. Make up your own.

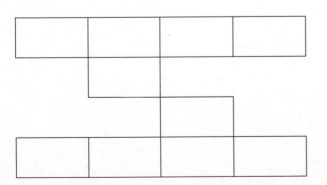

4.

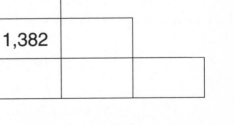

1,382

HOME LINK
1·3

Place-Value Practice

> **Family Note** In the last lesson, children learned how to use a number grid and how to solve number-grid puzzles. The **Try This** problems below give children more practice with what they have learned. For information about number grids and number-grid puzzles, see pages 7–9 in the *Student Reference Book*.
>
> *Please return this Home Link to school tomorrow. Also bring a clean sock tomorrow to use as an eraser with your slate.*
>
> **SRB**
> 7–9
> 18 19

1. Have someone at home tell you a four-digit number to write down.

   a. Write the the number. _____

   b. Circle the digit in the thousands place.

   c. Put an X through the digit in the tens place.

   d. Underline the digit in the ones place.

2. Write the number that is 100 more than your number in Problem 1. _____

3. Write the number that is 100 less than your number in Problem 1. _____

Try This

Use the filled-in grid on page 7 of your *Student Reference Book* to help.

4.

898	

5.

1,054	

Practice

Solve.

Unit

6. $4 + 5 =$ _____

7. _____ $= 9 - 5$

8. _____ $= 5 + 4$

9. $9 - 4 =$ _____

HOME LINK 1·4

Telling Time

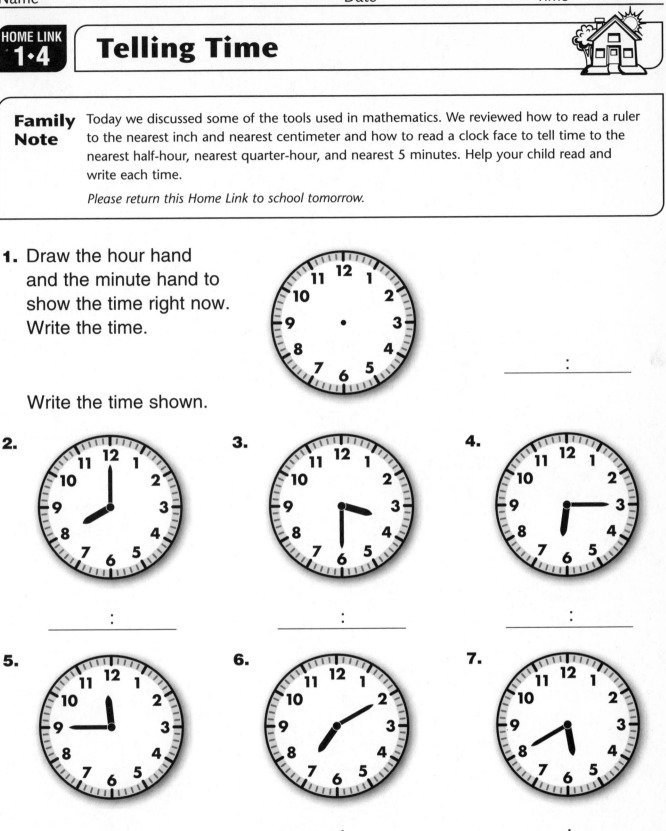

1. Draw the hour hand and the minute hand to show the time right now. Write the time.

         _____ : _____

Write the time shown.

2.   _____ : _____

3.   _____ : _____

4.   _____ : _____

5.   _____ : _____

6.   _____ : _____

7.   _____ : _____

8. Show someone at home how you solved the hardest problem on this page.

HOME LINK 1·5

How Much TV Did They Watch?

Family Note You can find information about tally charts on pages 76–78 in the *Student Reference Book.* You can find information about the minimum, maximum, range, mode, and median of a set of data on pages 79 and 81.

Please return this Home Link to school tomorrow.

SRB 76–78 79 81

Paul asked some of his classmates how many hours they watched television over the weekend. His classmates reported the following number of hours:

1 hour	3 hours	1 hour	5 hours	0 hours	2 hours
4 hours	3 hours	2 hours	3 hours	3 hours	

1. Make a tally chart for the data.

Time Spent Watching TV	
Hours	**Number of Children**
0	
1	
2	
3	
4	
5	

2. What was the least (minimum) number of hours watched? _____ hours

3. What was the greatest (maximum) number of hours watched? _____ hours

4. What is the range for the data? _____ hours (Remember that *range* is the difference between the greatest number and the least number.)

5. What is the mode for the data? _____ hours (Remember that the *mode* is the number that occurs most often.)

6. What is the median for the data? _____ hours (Remember that the *median* is the number in the middle.)

HOME LINK 1·6 | **Name-Collection Boxes**

Family Note You can find an explanation of name-collection boxes on pages 14 and 15 in the *Student Reference Book.*

Please return this Home Link to school tomorrow.

1. Write at least 10 names for the number 18 in the name-collection box. Then explain to someone at home how the box works. Have that person add another name for 18.

18

2. Three of the names do not belong in this box. Cross them out. Then write the name of the box on the tag.

~~HHT HHT~~   one dozen
7 + 5
number of months in 1 year
15 − 3   10 + 2
18 − 4   9 − 3

3. Make up a problem like Problem 2. Choose a name for the box but do not write it on the tag. Write 4 names for the number and 2 names that are not names for the number.

To check if the problem makes sense, ask someone at home to tell you which 2 names do not belong in the box. Have that person write the name of the box on the tag.

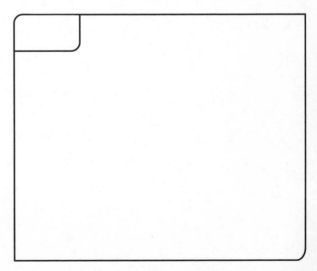

15

HOME LINK 1·7

Likely and Unlikely Events

Family Note During the next two weeks, please help your child find and cut out items in newspapers and magazines that discuss events that might or might not happen. Have your child bring these items to school to share with the class.

Please return this Home Link to school tomorrow.

SRB
92

For the next two weeks, look for items in newspapers and magazines that tell about events that **might** or **might not** happen. Get permission to cut them out and bring them to school. You might look for items like the following:

◆ a weather forecast (What are the chances that it will rain tomorrow?)

◆ the sports page (Which team is favored to win the baseball game?)

◆ a news story (What are the chances that people will explore distant planets in the next 20 years?)

Tell whether each event below is sure to happen, sure not to happen, or may happen, but not sure. Circle the answer.

1. You will grow taller next year.

sure to happen   sure not to happen   may happen, but not sure

2. You will live to be 200 years old.

sure to happen   sure not to happen   may happen, but not sure

3. You will watch TV next Saturday.

sure to happen   sure not to happen   may happen, but not sure

4. You will travel to the moon.

sure to happen   sure not to happen   may happen, but not sure

---

Practice

Solve.

5. $3 + 4 =$ _____

6. _____ $= 7 - 4$

7. _____ $= 7 - 3$

8. $4 + 3 =$ _____

Unit

17

HOME LINK 1·8 Finding Differences

Family Note It is not expected that your child knows how to use a traditional method of subtraction to solve these problems. Formal methods will be covered in the next unit. You can find an explanation of how to find differences on a number grid on page 8 in the *Student Reference Book*.

Please return this Home Link to school tomorrow.

SRB
8

1. Fill in the numbers on the number grid below.

	132								
									150
		154							
					177				

Use the number grid above to help you answer the following questions.

2. Which is more, 154 or 131? _____ How much more? _____

3. Which is less, 177 or 148? _____ How much less? _____

4. The difference between 180 and 158 is _____.

| **Try This** |

5. Explain how you found your answer in Problem 4.

_____

| **Practice** |

Solve.

6. $13 = 7 +$ _____

7. $13 = 6 +$ _____

8. $6 =$ _____ $- 7$

9. $7 =$ _____ $- 6$

Unit

19

HOME LINK 1·9

Large and Small Numbers

Family Note We have been reviewing place-value concepts in this lesson. For more information about place value, see pages 18 and 19 in the *Student Reference Book*.

Please return this Home Link to school tomorrow.

SRB 18–20

You will need a die or a deck of cards numbered from 0–9, or slips of paper numbered 0–9.

1. Roll a die 4 times (or draw 4 cards).

   a. Record the digit for each roll (or each card) in a blank.

   _____ _____ _____ _____

   b. Make the largest 4-digit number you can using these digits.

   _____, _____ _____ _____

   c. Make the smallest 4-digit number you can using these digits. The number may not begin with a zero.

   _____, _____ _____ _____

2. Roll a die 5 times (or draw 5 cards).

   a. Record the digit for each roll (or each card) in a blank.

   _____ _____ _____ _____ _____

   b. Make the largest 5-digit number you can using these digits.

   _____ _____, _____ _____ _____

   c. Make the smallest 5-digit number you can using these digits. The number may not begin with a zero.

   _____ _____, _____ _____ _____

Practice

Unit

Solve.

3. $8 = $ _____ $+ 5$

4. $8 = 5 +$ _____

5. _____ $= 8 - 3$

6. _____ $= 8 - 5$

21

HOME LINK 1·10 Ad Hunt

> **Family Note** The children have been working on dollars-and-cents notation (for example, $4.95). Help your child locate ads that clearly show prices.
>
> *Please return this Home Link to school tomorrow.*

1. Cut out four small advertisements from newspapers or magazines. Each ad must show the price of an item.

2. Put the ads in order from the least expensive item to the most expensive item.

3. Tape or glue your four ads in order on this page.

4. Bring extra ads to school to add to the Numbers All Around Museum.

Practice

Unit

5. Solve.

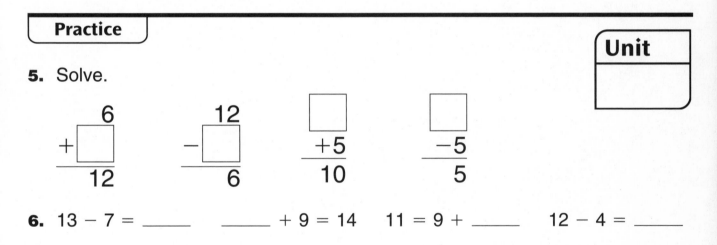

6. $13 - 7 =$ _____   _____ $+ 9 = 14$ $11 = 9 +$ _____ $12 - 4 =$ _____

HOME LINK 1·11 **Shopping in the Newspaper**

Family Note In this activity, your child will be looking for at least five different items to buy with $100. If any money is left over, your child can find something else to buy. If your child buys something in quantity (for example, 4 CDs), list each item and price on a separate line.

Please return this Home Link to school tomorrow.

SRB 191 193 194

1. Pretend that you have $100 to spend. Have someone at home help you find ads for at least five different items that you can buy. List the items and their prices below. DO NOT CALCULATE your total. Instead, estimate the total. You do not need to spend exactly $100.

Item	Actual Price	Estimated Price
CD	$15.75	$16

2. Explain to someone at home how you estimated the total price of your items.

Practice

Solve.

Unit

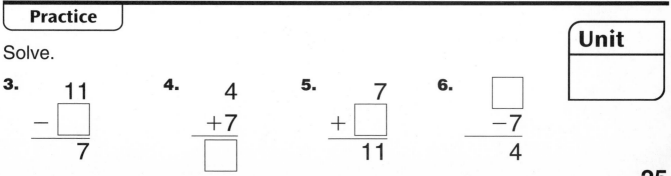

3.
```
   11
 − □
 ----
   7
```

4.
```
   4
 +7
 ----
  □
```

5.
```
   7
 +□
 ----
  11
```

6.
```
   □
 −7
 ----
   4
```

25

HOME LINK 1·12 Frames-and-Arrows

Family Note You can find information about Frames–and–Arrows diagrams on pages 200 and 201 in the *Student Reference Book*.

Please return this Home Link to school tomorrow.

SRB 200 201

Show someone at home how to complete these Frames-and-Arrows diagrams.

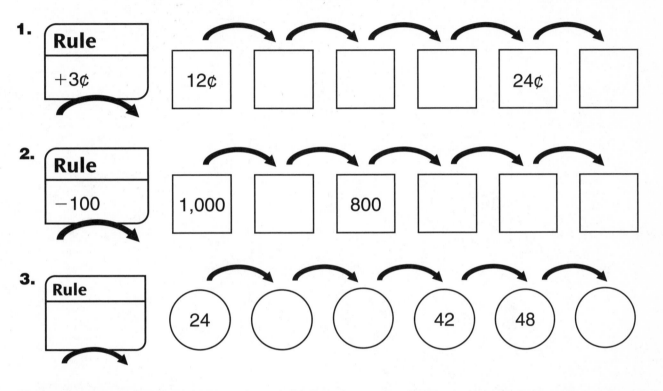

1. **Rule** +3¢

   12¢ | | | | 24¢ |

2. **Rule** −100

   1,000 | | 800 | | |

3. **Rule**

   24 | | | 42 | 48 |

Practice

Write each amount in dollars-and-cents notation.

4. $1 Q D N N P = $_____

5. D D Q N P D Q P = $_____

6. $10 $1 $1 N P = $_____

7. Draw coins to show $0.89 in at least two different ways.

_____

_____

27

HOME LINK 1·13 | **Time Practice**

Family Note   Your child has been learning about elapsed time in this lesson.
Please return this Home Link to school tomorrow.

SRB
174

Pretend you are setting your watch. Draw the hour hand and minute hand on the clock face to show the time. Use a real watch or clock to help you.

1. a. Show a quarter to 6.

b. Show the time 2 hours and 15 minutes later.

2. a. Show half-past 8.

b. Show the time 4 hours and 20 minutes earlier.

3. a. Show 25 minutes past 11.

b. Show the time 3 hours and 40 minutes later.

Practice

Solve.

4. $4 + \underline{\hspace{1cm}} = 8$ **5.** $\underline{\hspace{1cm}} = 8 - 4$ **6.** $14 = \underline{\hspace{1cm}} + 7$ **7.** $14 - \underline{\hspace{1cm}} = 7$

HOME LINK 1·14

Unit 2: Family Letter

Adding and Subtracting Whole Numbers

Unit 2 will focus on addition and subtraction of whole numbers, emphasizing problem-solving strategies and computational skills. In *Second Grade Everyday Mathematics*, children used shortcuts, fact families, Fact Triangles, and games to help them learn basic addition and subtraction facts. Such devices will continue to be used in third grade. Knowledge of the basic facts and their extensions is important. Knowing that 6 + 8 = 14, for example, makes it easy to solve such problems as 56 + 8 = ? and 60 + 80 = ? Later, knowing that 5 × 6 = 30 will make it easy to solve 5 × 60 = ?, 50 × 60 = ?, and so on.

In Unit 2, children will learn new methods for solving addition and subtraction problems. *Everyday Mathematics* encourages children to choose from any of these methods or to invent their own computation methods. When children create—and share—their own ways of doing operations instead of simply learning one method, they begin to realize that any problem can be solved in more than one way. They are more willing and able to take risks, think logically, and reason analytically.

Blair Chewning, a teacher in Richmond, Virginia, gave her Everyday Mathematics *students this problem to solve. Here are just a few of the strategies her students used.*

Jill needs to earn $45.00 for a class trip. She earns $2 per day on Mondays, Tuesdays, and Wednesdays. She earns $3 each day on Thursdays, Fridays, and Saturdays. She does not work on Sundays. How many weeks will it take her to earn $45?

(3 weeks)

2 9 6 $6.00
3 6 9 $9.00

$6.00
+ $9.00

15 3 weeks $15.00   15 1 week
+ 30   + 15 2 weeks
45   30

2 1   2 1   2 1   3 1   3 1   3 1
Mon   Tu   Wen   Th   Fr.   St.
2   2   2   3   3   3
4   4   4   6   6   6
6   6   6   9   9   9

15 + 15 + 15 (3 wks.)

3 weeks $6.00 $9.00 $45.00
3 × $15.00 = $45.00

2 + 2 + 2   3 + 3 + 3 = 15.00
2 + 2 + 2   3 + 3 + 3 = 30.00
2 + 2 + 2   3 + 3 + 3 = 45.00
(3 weeks.)

3 days 6   - 6
3 days more 9   9
= 15 a week
   1   1
   15   15
   15   15
   30   15
     $45.00

Answer
I t will take her 3 weeks.

Finally, Unit 2 introduces another yearlong project—the National High/Low Temperatures Project. Children will calculate, record, and graph differences in temperatures from cities around the United States.

Vocabulary

ballpark estimate A rough estimate. A ballpark estimate can be used when you don't need an exact answer or to check if an answer makes sense.

fact family A collection of 4 related addition and subtraction facts, or multiplication and division facts, relating 3 numbers.

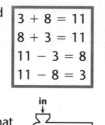

$$3 + 8 = 11$$
$$8 + 3 = 11$$
$$11 - 3 = 8$$
$$11 - 8 = 3$$

function machine In *Everyday Mathematics,* an imaginary machine that processes numbers and pairs them with output numbers according to a set rule. A number (input) is put into the machine and is transformed into a second number (output) through the application of the rule.

"What's My Rule?" problems A problem in which number pairs are related to each other according to the same rule. Sometimes the rule and one number in each pair are given, and the other number is to be found. Sometimes the pairs are given and the rule is to be found.

in	out
3	8
5	10
8	13
10	15
16	21

number family Same as a fact family.

number model A number sentence that shows how the parts of a number story are related. For example, 5 + 8 = 13 models the number story: *5 children skating. 8 children playing ball. How many children in all?*

parts-and-total diagram A diagram used to represent problems in which two or more quantities are combined to form a total quantity. Sometimes

the parts are known and the total is unknown. Other times the total and one or more parts are known, but one part is unknown.

For example, the parts-and-total diagram here represents this number story: *Leo baked 24 cookies. Nina baked 26 cookies. How many cookies in all?*

Total	
50	
Part	Part
24	**26**

change diagram A diagram used to represent addition or subtraction problems in which a given quantity is increased or decreased. The diagram includes the starting quantity, the ending quantity, and the amount of the change.

For example, the change diagram here represents this subtraction problem: *Rita had $28 in her wallet. She spent $12 at the store. How much money is in Rita's wallet now?*

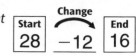

comparison diagram A diagram used to represent problems in which two quantities are given and then compared to find how much more or less one quantity is than the other.

For example, the comparison diagram here represents this problem: *34 children ride the bus to school. 12 children walk to school. How many more children ride the bus?*

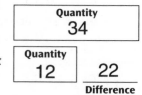

unit box In *Everyday Mathematics,* a box displaying the unit for numbers in the problems at hand.

Unit box

32

Math Tools

Your child will be using **Fact Triangles** to practice and review addition and subtraction facts. Fact Triangles are a new and improved version of flash cards; the addition and subtraction facts shown are made from the same three numbers, and this helps your child understand the relationships among those facts.

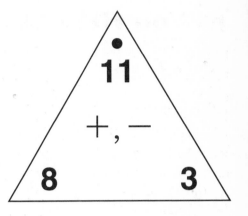

Do-Anytime Activities

To work with your child on the concepts taught in this unit and in the previous unit, try these interesting and rewarding activities:

1. Review addition and subtraction facts. Make +,− Fact Triangles for facts that your child needs to practice.

2. Practice addition and subtraction fact extensions. *For example:*

6 + 7 = 13	13 − 7 = 6
60 + 70 = 130	23 − 7 = 16
600 + 700 = 1,300	83 − 7 = 76

3. When your child adds or subtracts multidigit numbers, talk about the strategy that works best. Try not to impose the strategy that works best for you! Here are some problems to try:

267 + 743 = _____

794 − 554 = _____

_____ = 851 + 697

840 − 694 = _____

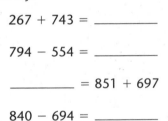

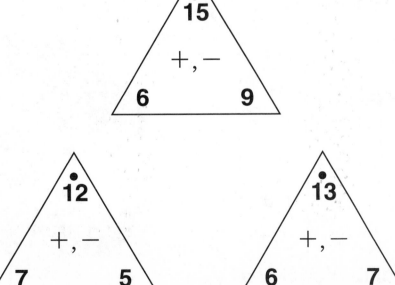

As You Help Your Child with Homework

As your child brings home assignments, you may want to go over the instructions together, clarifying them as necessary. The answers listed below will guide you through this unit's Home Links.

Home Link 2·1

1. $9 + 6 = 15$; $6 + 9 = 15$; $15 - 9 = 6$; $15 - 6 = 9$

2. $25 + 50 = 75$; $50 + 25 = 75$; $75 - 25 = 50$; $75 - 50 = 25$ **3.** Answers vary.

4. 10   **5.** 12   **6.** 4   **7.** 10

Home Link 2·2

1. 16; 26; 76; 106   **2.** 12; 22; 62; 282

3. 8; 28; 58; 98   **4.** 5; 15; 115; 475

5. 13; 130; 1,300; 13,000

Home Link 2·3

1.

in	out
14	7
7	0
12	5
15	8
10	3
21	14 Answers vary.

2.

in	out
7	16
9	18
37	46
77	86
49	58

3.

in	out
70	100
20	50
30	60
90	120
50	80

Rule: Add 30

Answers vary.

Home Link 2·4

1. 55 minutes; $25 + 30 = 55$

2. 700 cans; $300 + 400 = 700$

Home Link 2·5

1. $9; $25 - 16 = 9$ **2.** $49; $35 + 14 = 49$
or $16 + 9 = 25$

Home Link 2·6

1. $29; $42 - 13 = 29$ **2.** 9 days; $28 - 19 = 9$
or $13 + 29 = 42$ or $19 + 9 = 28$

3. 15 children; $40 - 25 = 15$

Home Link 2·7

1. 337   **2.** 339   **3.** 562
4. 574   **5.** 627   **6.** 1,214

Home Link 2·8

1. 194   **2.** 202   **3.** 122
4. 206   **5.** 439   **6.** 487

Home Link 2·9

1. 38   **2.** 213   **3.** 40
4. 70   **5.** 915   **6.** 55; $18 + 15 + 22 = 55$
7. 19; $17 + 22 + 19 = 58$

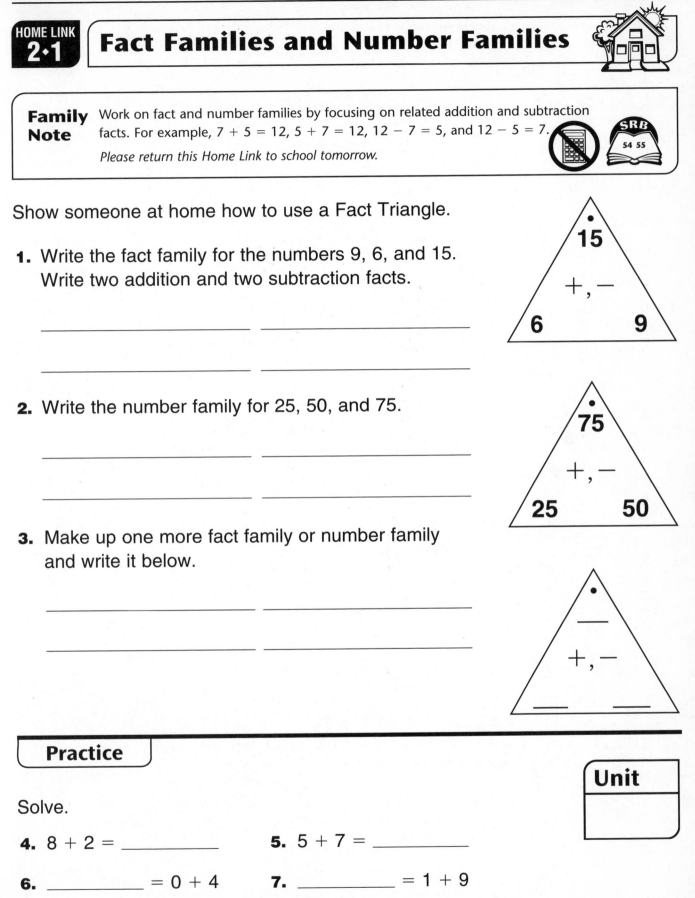

HOME LINK 2·1

Fact Families and Number Families

Family Note Work on fact and number families by focusing on related addition and subtraction facts. For example, 7 + 5 = 12, 5 + 7 = 12, 12 − 7 = 5, and 12 − 5 = 7.

Please return this Home Link to school tomorrow.

SRB 54 55

Show someone at home how to use a Fact Triangle.

1. Write the fact family for the numbers 9, 6, and 15. Write two addition and two subtraction facts.

_____ _____

_____ _____

15

+, −

6   9

2. Write the number family for 25, 50, and 75.

_____ _____

_____ _____

75

+, −

25   50

3. Make up one more fact family or number family and write it below.

_____ _____

_____ _____

___

+, −

___ ___

Practice

Solve.

4. 8 + 2 = _____

5. 5 + 7 = _____

6. _____ = 0 + 4

7. _____ = 1 + 9

Unit

HOME LINK 2·2   **Fact Extensions**

Family Note   Knowing basic facts, such as 6 + 7 = 13, makes it easy to solve similar problems with larger numbers, such as 60 + 70 = 130. Help your child think of more fact extensions to complete this Home Link.

Please return this Home Link to school tomorrow.

SRB
50 51

Write the answer for each problem.

1. I know: 9
                 + 7

This helps me know: 19   69   99
                      + 7   + 7   + 7

2. I know: 8
                 + 4

This helps me know: 18   58   278
                      + 4   + 4   + 4

3. I know: 15
                 − 7

This helps me know: 35   65   105
                      − 7   − 7   − 7

4. I know: 13
                 − 8

This helps me know: 23   123   483
                      − 8   − 8   − 8

5. I know: 6
                 + 7

This helps me know: 60   600   6,000
                     + 70   + 700   + 7,000

Make up another set of fact extensions.

6. I know: ▢   This helps me know: ▢   ▢   ▢
         ▢   ▢   ▢   ▢

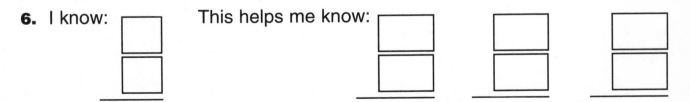

HOME LINK 2·3 "What's My Rule?"

Family Note You can find an explanation of function machines and "What's My Rule?" tables on pages 202–204 in the *Student Reference Book*. Ask your child to explain how they work. Help your child fill in all the missing parts for these problems.

Please return this Home Link to school tomorrow.

SRB
202 204

Practice facts and fact extensions. Complete the "What's My Rule?" problems. Make up problems of your own for the last table.

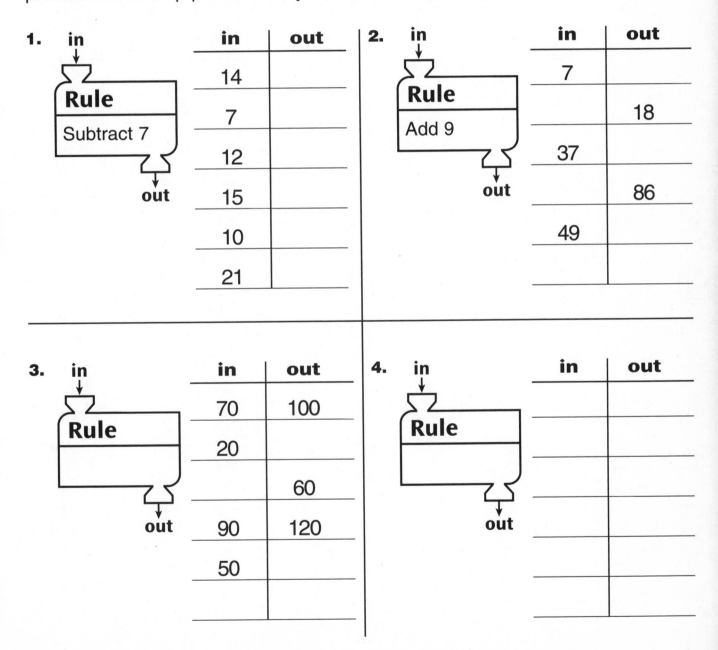

1.

Rule: Subtract 7

in	out
14	
7	
12	
15	
10	
21	

2.

Rule: Add 9

in	out
7	
	18
37	
	86
49	

3.

Rule:

in	out
70	100
20	
	60
90	120
50	

4.

Rule:

in	out

HOME LINK
2·4

Parts-and-Total Number Stories

Family Note   Today your child learned about a diagram that helps organize the information in a number story. We call it a *parts-and-total diagram*. For more information, see pages 256 and 257 in the *Student Reference Book*.

Please return this Home Link to school tomorrow.

SRB
256 257

For each problem, write *?* for the number you want to find. Write the numbers you know in the diagram. Then write the answer and a number model. Finally, write how you know that each answer makes sense.

1. Marisa read her book for 25 minutes on Monday and 30 minutes on Tuesday. How many minutes in all did she read?

   Answer the question: _____
   (unit)

   Number model: _____

   Check: How do you know your answer makes sense?

   _____

   _____

Total	
Part	**Part**

2. The second graders collected 300 cans to recycle. The third graders collected 400 cans. What was the total number of cans they collected?

   Answer the question: _____
   (unit)

   Number model: _____

   Check: How do you know your answer makes sense?

   _____

   _____

Total	
Part	**Part**

41

HOME LINK 2·5

Change Number Stories

Family Note   Today your child learned about another diagram that helps organize the information in a number story. It is called a *change diagram.* For more information, see pages 254 and 255 in the *Student Reference Book.*

Please return this Home Link to school tomorrow.

SRB
254 255

For each number story, write ? in the diagram for the number you want to find. Write the numbers you know in the change diagram. Then, write the answer and a number model. Finally, write how you know that each answer makes sense.

1. Marcus had $25 in his wallet. He spent $16 at the store. How much money was in Marcus's wallet then?

    Change

    Start ____ **End**

    Answer the question: _____
    (unit)

    Number model: _____

    Check: How do you know your answer makes sense?

    _____

    _____

2. Jasmine had $35. She earned $14 helping her neighbors. How much money did she have then?

    Change

    Start ____ **End**

    Answer the question: _____
    (unit)

    Number model: _____

    Check: How do you know your answer makes sense?

    _____

    _____

Comparison Number Stories

Family Note Today your child learned about a comparison diagram. It helps organize information in a number story. To read more, see page 258 in the *Student Reference Book*.

Please return this Home Link to school tomorrow.

SRB
258

Write ? in the diagram for the number you want to find. Write the numbers you know in the diagram. Then write the answer and a number model. Tell someone at home how you know that your answers make sense.

1. Jenna has $42. Her brother has $13. How much more money does Jenna have?

Answer the question: _____
(unit)

Number model:

Quantity

Quantity	
	_____

Difference

---

2. There are 28 days until Pat's birthday and 19 days until Ramon's birthday. How many more days does Pat have to wait than Ramon?

Answer the question: _____
(unit)

Number model:

Quantity

Quantity	
	_____

Difference

---

3. There are 25 children in the soccer club and 40 children in the science club. How many fewer children are in the soccer club?

Answer the question: _____
(unit)

Number model:

Quantity

Quantity	
	_____

Difference

HOME LINK 2·7 — The Partial-Sums Addition Method

Solve each addition problem. You may want to use the partial-sums method. Use a ballpark estimate to check that your answer makes sense. Write a number model to show your estimate.

1. Ballpark estimate:	2. Ballpark estimate:	3. Ballpark estimate:
_____	_____	_____
100s 10s 1s 2 4 5 + 9 2	124 + 215	245 + 317
4. Ballpark estimate:	**5.** Ballpark estimate:	**6.** Ballpark estimate:
_____	_____	_____
366 + 208	459 + 168	769 + 445

HOME LINK 2·8 **Subtraction Methods**

Family Note   Over the past 2 days, your child practiced subtracting two 3-digit numbers using the counting-up method and the trade-first method. For more information, see pages 60, 61, and 63 in the *Student Reference Book*.

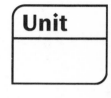

Please return this Home Link to school tomorrow.

Fill in the unit. Solve the problems. You may use any method you wish. Use a ballpark estimate to check that your answer makes sense. Write a number model for your estimate. On the back of this Home Link, explain how you solved one of the problems.

Unit

---

1. Ballpark estimate:

_____

$$\begin{array}{r} 468 \\ -274 \\ \hline \end{array}$$

2. Ballpark estimate:

_____

$$\begin{array}{r} 531 \\ -329 \\ \hline \end{array}$$

3. Ballpark estimate:

_____

$$\begin{array}{r} 331 \\ -209 \\ \hline \end{array}$$

---

4. Ballpark estimate:

_____

$$\begin{array}{r} 653 \\ -447 \\ \hline \end{array}$$

5. Ballpark estimate:

_____

$$\begin{array}{r} 925 \\ -486 \\ \hline \end{array}$$

6. Ballpark estimate:

_____

$$\begin{array}{r} 724 \\ -237 \\ \hline \end{array}$$

49

HOME LINK 2·9

Three or More Addends

> **Family Note** This Home Link provides practice in looking for combinations that make addition easier. Guide your child to look for combinations that add up to 10, 20, 30, 40 and so on. Then add the rest of the numbers.
>
> *Please return this Home Link to school tomorrow.*

Remember that when you add:

♦ The numbers can be in any order.

♦ Some combinations make the addition easier.

Add. Write the numbers in the order you added them. Tell someone at home why you added the numbers in that order.

Example:

$5 + 17 + 25 + 3 =$ __50__

I added in this order:

$$5 + 25 + 17 + 3$$

---

1. $6 + 18 + 14 =$ _____

I added in this order:

_____

---

2. $125 + 13 + 75 =$ _____

I added in this order:

_____

---

3. $15 + 6 + 14 + 5 =$ _____

I added in this order:

_____

---

4. $33 + 22 + 8 + 7 =$ _____

I added in this order:

_____

---

5. $150 + 215 + 300 + 50 + 200 =$ _____

I added in this order:

_____

51

HOME LINK 2·9

Three or More Addends *continued*

Solve these number stories.

6. Nico's baby brother has a basket of wooden blocks. 18 blocks are red, 15 are blue, and 22 are yellow. How many red, blue, and yellow blocks are in the basket?

Answer the question: _____ blocks

Number model: _____

Total		
Part	**Part**	**Part**

7. Marianna has 3 days to read a 58-page book. She read 17 pages on Monday and 22 pages on Tuesday. How many more pages does she need to read to finish the book?

Answer the question: _____ pages

Number model: _____

Total		
Part	**Part**	**Part**

8. Make up a number story with three or more addends.

_____

_____

_____

_____

_____

Answer the question: _____
                            (unit)

Number model: _____

Check: Does my answer make sense?

HOME LINK 2·10 | Unit 3: Family Letter

Linear Measures and Area

In Unit 3, children will develop their measurement sense by measuring lengths with standard units—in both the **U.S. customary system** and the **metric system.**

Children will practice reading a ruler to the nearest inch, nearest $\frac{1}{2}$ inch, nearest $\frac{1}{4}$ inch, and nearest centimeter as they measure a variety of objects, including parts of their own bodies, such as their hand spans, wrists, necks, and heights. In addition to the inch and centimeter, children will also measure with other standard units, such as the foot, yard, and meter. Children will begin to use certain body measures or the lengths of some everyday objects as **personal references** to estimate the lengths of other objects or distances. For example, a sheet of notebook paper that is about 1 foot long can help children estimate the length of a room in feet.

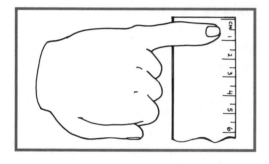

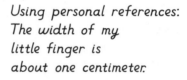

Using personal references: The width of my little finger is about one centimeter.

The concept of **perimeter** is also investigated in this unit. Children will use straws and twist-ties to build **polygons,** or 2-dimensional figures having connected sides. Then children will measure the distance around each polygon to find the perimeter.

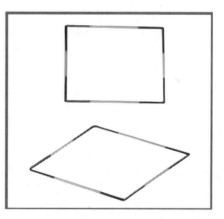

Children will also discover the meaning of **area** by tiling small rectangles with blocks and counting how many blocks cover the rectangles. Children see how to calculate area by tiling larger surfaces, such as tabletops and floors, with square feet and square yards.

In the last part of this unit, children will explore the **circumference** and **diameter** of circles. They will learn the *about 3 times* rule—that the circumference of a circle is a little more than 3 times the length of its diameter.

Please keep this Family Letter for reference as your child works through Unit 3.

Vocabulary

Important terms in Unit 3:

unit An agreed-upon unit of measure, for example foot, pound, gallon, meter, kilogram, liter.

length The distance between two points.

U.S. customary system The measurement system used in the United States. For example, inches, feet, yards, and miles are used to measure length.

metric system of measurement A measurement system based on the base-ten numeration system. It is used in most countries around the world. For example, millimeters, centimeters, meters, and kilometers are used to measure length.

benchmark A well-known count or measure that can be used to check whether other counts, measures, or estimates make sense. For example, a benchmark for land area is that a football field is about one acre. A benchmark for length is that the width of a man's thumb is about one inch. Benchmarks are also called *personal-measurement references*.

perimeter The distance around the boundary of a 2-dimensional shape. The perimeter of a circle is called its *circumference*. A formula for the perimeter P of a rectangle with length l and width w is $P = 2 \times (l + w)$.

circumference The perimeter of a circle.

diameter A line segment that passes through the center of a circle or sphere. The length of such a segment.

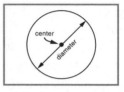

polygon A 2-dimensional figure formed by 3 or more line segments (sides) that meet only at their endpoints (vertices) to make a closed path. The line segments of a polygon may not cross.

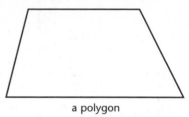

a polygon

tiling The covering of a surface with shapes so that there are no gaps or overlaps.

area The amount of surface inside a 2-dimensional figure. Area is measured in square units, such as square inches or square centimeters.

square unit A unit used to measure area; a square that measures 1 inch, 1 centimeter, 1 yard, or 1 other standard measure of length on each side.

1 square centimeter   1 square inch

Do-Anytime Activities

To work with your child on the concepts taught in this unit and in previous units, try these interesting and rewarding activities:

1. Encourage your child to find some personal references for making several measurements of length at home.

2. Practice using the personal references by *estimating* some lengths, and then practice using a ruler by *measuring* the actual lengths.

3. Practice finding perimeters of objects and circumferences of circular objects around your home.

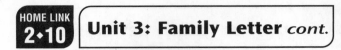
As You Help Your Child with Homework

As your child brings home assignments, you may want to go over the instructions together, clarifying them as necessary. The answers listed below will guide you through this unit's Home Links.

Home Link 3·4

2. perimeter of polygon A = 20 cm

perimeter of polygon B = 20 cm

3. a. 12 ft   **3. b.** 60 in.

Home Link 3·5

1. 6   **2.** 2   **3.** 4   **4.** 3

5. 3   **6.** 95   **7.** 62

Home Link 3·7

1. Area = 24 square units   **2.** Area = 27 square units

Sample answer: Sample answer:

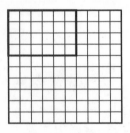

3. This is a 2-by-6 rectangle. Area = 12 square units

4. This is a 5-by-4 rectangle. Area = 20 square units

5. 307   **6.** 119

Home Link 3·8

1. 80 tiles   **2.** $160

3.   **4.** 30 plants

5. 489

6. 673

7. 307

Building Skills through Games

In Unit 3, your child will practice addition skills by playing the following games. For detailed instructions, see the *Student Reference Book.*

Addition Top-It

Each player turns over two cards and calls out their sum. The player with the higher sum then takes all the cards from that round.

Subtraction Top-It

Each player turns over two cards and calls out their difference. The player with the larger difference then takes all the cards from that round.

HOME LINK
3·1

Measurements at Home

> **Family Note** Help your child find labels, pictures, and descriptions that contain measurements. If possible, collect them in an envelope or folder so that your child can bring them to school tomorrow, along with this Home Link.
>
> *Please return this Home Link to school tomorrow.*

1. Find items with measurements on them. Look at boxes and cans.
   List the items and their measurements.

Item	Measurement
milk carton	*1 quart*

2. Find pictures and ads that show measurements. Look in
   newspapers, magazines, or catalogs. Ask an adult if you can cut out
   some examples and bring them to school.

Practice

Write these problems on the back of this page. Write a number model
for your ballpark estimate. Use any method you wish to solve each problem.
Show your work.

3. 259 + 432 = _____

4. 542 − 387 = _____

HOME LINK 3·2   Body Measures

Family Note   Help your child measure an adult at home. Use a tape measure if you have one, or use a piece of string. Mark lengths on the string with a pen, and then measure the string with a ruler.

Please return this Home Link to school tomorrow.

Measure an adult at home to the nearest $\frac{1}{2}$ inch. Fill in the information below:

Name of adult: _____

Height: about _____ inches

Length of shoe: about _____ inches

Around neck: about _____ inches

Around wrist: about _____ inches

Distance from waist to floor:

about _____ inches

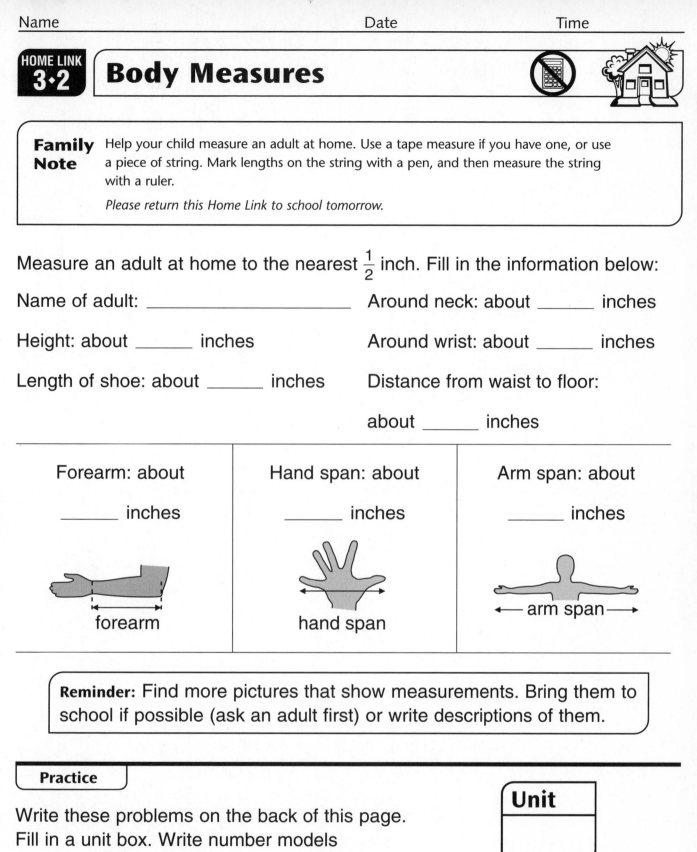

Forearm: about	Hand span: about	Arm span: about
_____ inches	_____ inches	_____ inches
forearm	hand span	arm span

Reminder: Find more pictures that show measurements. Bring them to school if possible (ask an adult first) or write descriptions of them.

Practice

Write these problems on the back of this page.
Fill in a unit box. Write number models
for your ballpark estimates. Show your work.

Unit

1. 83 − 25 = _____   **2.** _____ = 35 + 47   **3.** 58 + 89 = _____

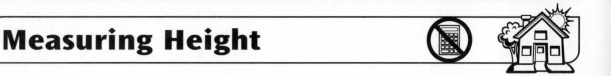

HOME LINK 3·3 | **Measuring Height**

> **Family Note**
>
> Measuring the height of the ceiling is easiest with such tools as a yardstick, a carpenter's ruler, or a metal tape measure. Another way is to attach a string to the handle of a broom and raise it to the ceiling. Have the string extend from the ceiling to the floor, cut the string to that length, and then measure the string with a ruler.
>
> *Please return this Home Link to school tomorrow.*

Work with someone at home.

1. Measure the height of the ceiling in your room.

   The ceiling in my room is about _____ feet high.

2. Measure the height of a table.

   The table is between _____ and _____ feet high.

3. About how many tables could you stack in your room, one on top of the other?

   about _____ tables

4. Draw a picture on the back of this page to show how the tables might look stacked in your room.

Practice

Write these problems on the back of this page. Draw and fill in a unit box. Write a number model for your ballpark estimate. Use any method you wish to solve each problem. Show your work.

5. _____ = 63 + 28   6. 149 − 76 = _____

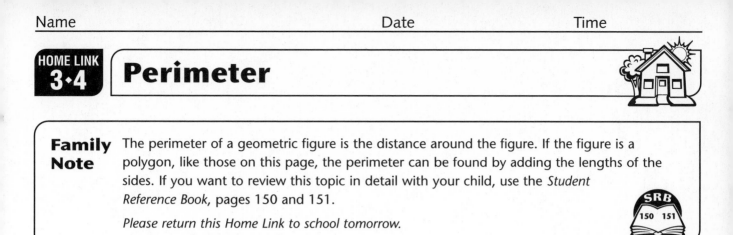

HOME LINK 3·4   Perimeter

1. Estimate: Which has the larger perimeter, polygon A or polygon B? _____

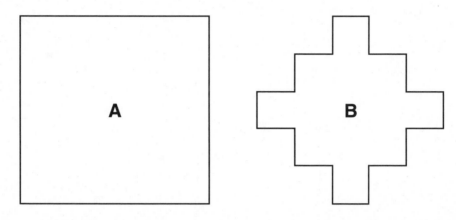

2. Check your estimate by measuring the perimeter of each polygon in centimeters. If you don't have a centimeter ruler, cut out the one at the bottom of the page.

perimeter of polygon A = _____ cm   perimeter of polygon B = _____ cm

3. What is the perimeter of each figure below?

a.

5 ft
3 ft
4 ft

b. each side 10 inches

perimeter = _____ ft   perimeter = _____ in.

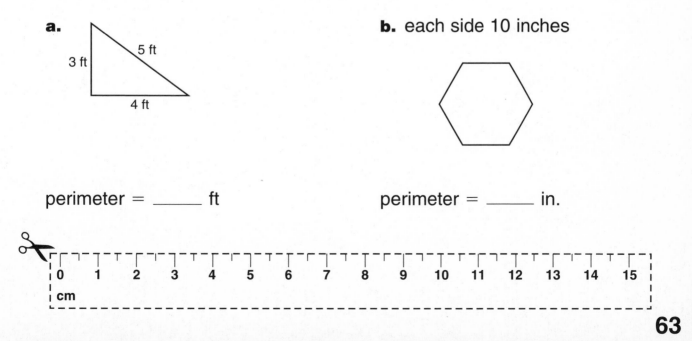

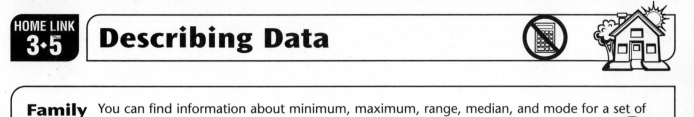

HOME LINK 3·5 Describing Data

Children in the Science Club collected pill bugs. The tally chart shows how many they collected. Use the data from the tally chart to complete a line plot.

Number of Pill bugs	Number of Collectors
0	
1	
2	///
3	~~////~~
4	
5	//
6	//

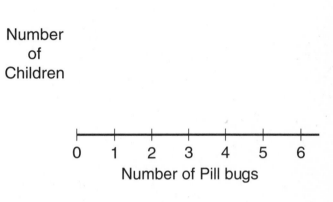

Number of Children

```
|----|----|----|----|----|----|----|
0    1    2    3    4    5    6
       Number of Pill bugs
```

Use the data to answer the questions.

1. What is the maximum (greatest) number of pill bugs found? _____ pill bugs

2. What is the minimum (least) number of pill bugs found? _____ pill bugs

3. What is the range for the data? _____ pill bugs

4. What is the median for the data? _____ pill bugs

5. What is the mode for the data? _____ pill bugs

Practice

Make ballpark estimates. Solve on the back of this paper.
Show your work.

Unit

6. $67 + 28 =$ _____

7. $33 + 29 =$ _____

HOME LINK 3·6 Room Perimeters

Family Note A personal measurement reference is something you know the measure of—for example, your height or ounces in a water bottle. Personal references can help you estimate measures that you don't know. A person's pace can be defined as the length of a step, measured from heel to heel or from toe to toe. It will be helpful for you to read about Personal Measurement References on pages 141, 142, 148, and 149 in the *Student Reference Book* with your child.

Please return this Home Link to school tomorrow.

SRB
141 142
148 149

Your pace is the length of one of your steps.

1. Find the perimeter, in paces, of your bedroom.
   Walk along each side and count the number of paces.

   The perimeter of my bedroom is about _____ paces.

2. Which room in your home has the largest perimeter? Use your estimating skills to help you decide.

   The _____ has the largest perimeter.

   Its perimeter is about _____ paces.

3. Draw this room on another sheet of paper.
   Plan to share your drawing with the class.

Practice

Write these problems on the back of this page. Fill in a unit box. Write a number model for your ballpark estimate. Use any method you wish to solve each problem. Show your work.

4. $38 + 9 =$ _____

5. $143 - 37 =$ _____

6. _____ $= 576 - 67$

Unit

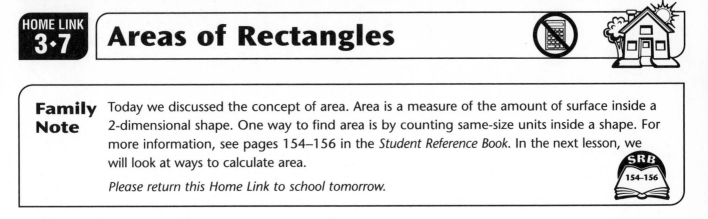

HOME LINK 3·7

Areas of Rectangles

Family Note Today we discussed the concept of area. Area is a measure of the amount of surface inside a 2-dimensional shape. One way to find area is by counting same-size units inside a shape. For more information, see pages 154–156 in the *Student Reference Book*. In the next lesson, we will look at ways to calculate area.

Please return this Home Link to school tomorrow.

SRB 154–156

Show someone at home how to find the area of each rectangle. Make a dot in each square as you count the squares inside the rectangle.

1. Draw a 4-by-6 rectangle on the grid.

2. Draw a 3-by-9 rectangle.

Fill in the blanks.

3.

This is a _____-by-_____ rectangle.

Area = _____ square units

4.

This is a _____-by-_____ rectangle.

Area = _____ square units

Practice

Write these problems on the back of this page. Fill in a unit box. Use any method you wish to solve each problem. Write a number model for your ballpark estimate. Show your work.

Unit

5. 571
  − 264

6. 805
  − 686

HOME LINK 3·8   Area

Family Note

Today we discussed area as an array, or diagram. An array is a rectangular arrangement of objects in rows and columns. Help your child draw an array of the tomato plants in Problem 3. Use that diagram to find the total number of plants.

Please return this Home Link to school tomorrow.

SRB
64 65

Mr. Li tiled his kitchen floor.
This is what the tiled floor looks like.

1. How many tiles did he use? _____ tiles

2. Each tile cost $2. How much did all the tiles

cost? $_____

3. Mrs. Li planted tomato plants. She planted
5 rows with 6 plants in each row. Draw a diagram
of her tomato plants.
Hint: You can show each plant with a large dot or an X.

4. How many tomato plants are there in all?

_____ plants

Practice

Write these problems on the back of this page. Fill in a unit box. Write a
number model for your ballpark estimate. Use any method you wish to solve
each problem. Show your work.

5. 548 − 59 = _____

6. _____ = 616 + 57

7. _____ = 571 − 264

Unit

HOME LINK 3·9 | Circumference and Diameter

Family Note

Today in school your child learned the definitions of *circumference* and *diameter*. Ask your child to explain them to you. Help your child find and measure circular objects, such as cups, plates, clocks, cans, and so on. The *about 3 times* circle rule says that the circumference of any circle, no matter what size, is about 3 times its diameter. It will be helpful for you to review pages 152 and 153 in the *Student Reference Book* with your child.

Please return this Home Link to school tomorrow.

SRB
152 153

Measure the diameters and circumferences of circular objects at home. Use a tape measure if you have one, or use a piece of string. Mark lengths on the string with your finger or a pen, and then measure the string. Record your measures in the chart below.

Does the *about 3 times* circle rule seem to work? Share the *about 3 times* rule with someone at home.

Diameter = 9 cm

Circumference = about 27 cm

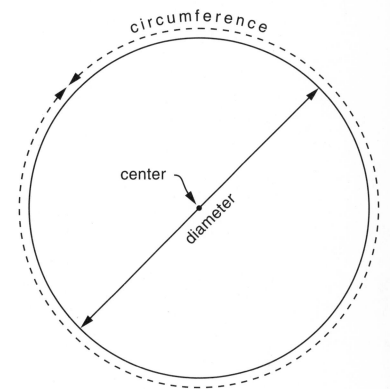

Object	Diameter	Circumference

HOME LINK 3·10

Unit 4: Family Letter

Multiplication and Division

Unit 4 focuses on the most common uses of multiplication and division—problems that involve equal sharing and equal grouping. In *Second Grade Everyday Mathematics*, children were exposed to multiplication and division number stories and multiplication and division facts. To solve multiplication and division number stories, children will refer to familiar strategies introduced in second grade:

◆ **Acting out problems using concrete objects, such as counters** (below)

$3 \times 4 = 12$

$2 \times 7 = 14$

◆ **Using diagrams to sort out quantities** (below)

children	pennies per child	pennies in all
4	?	28

◆ **Using number models to represent solution strategies** (below)

◆ **Representing problems with pictures and arrays** (below)

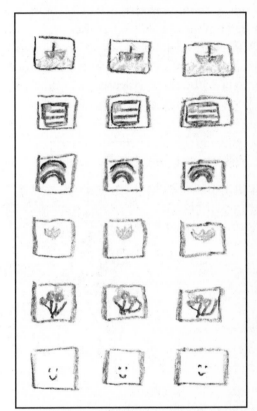

A sheet of stamps has 6 rows. Each row has 3 stamps. How many stamps are on a sheet?

$6 \times 3 = 18$

Problem:	Solution strategies:
Each child has 2 apples. There are 16 apples. How many children have apples?	$2 \times ? = 16$, or I know that $16 \div 2 = 8$. If there are 16 apples and each child has 2, then there must be 8 children.

Vocabulary

Important terms in Unit 4:

multiples of a number The product of the number and a counting number. For example, multiples of 2 are 2, 4, 6, and 8....

multiplication/division diagram In *Everyday Mathematics*, a diagram used to represent problems in which the total number of objects in several equal groups is being considered. The diagram has three parts: number of groups, number in each group, and total number. For example, the multiplication/division diagram here represents this number story: There are 3 boxes of crayons. Each box has 8 crayons. There are 24 crayons in all.

boxes	crayons per box	crayons
3	8	24

rectangular array A group of objects placed in rows and columns.

A 2-by-6 array of eggs

factor Each of the two or more numbers in a product.

> In **4 × 3 = 12**,
> **4** and **3** are the **factors**,
> and **12** is the **product**.

product The result of multiplying two numbers.

equal groups Sets with the same number of elements, such as tables with 4 legs, rows with 6 chairs, boxes of 100 paper clips, and so on.

dividend The number in division that is being divided.

divisor In division, the number that divides another number, the *dividend*.

quotient The result of division.

> In **28 ÷ 4 = 7,**
> **28** is the **dividend,**
> **4** is the **divisor,** and
> **7** is the **quotient.**

remainder An amount left over when one number is divided by another number. In the division number model $16 ÷ 3 \rightarrow 5$ R1, the remainder is 1.

square number The product of a number multiplied by itself; any number that can be represented by a square array of dots or objects. A square array has the same number of rows as columns.

> $3 × 3 = 9$
> The number 9 is a **square number.**

Building Skills through Games

In Unit 4, your child will practice division and multiplication by playing the following games. For detailed instructions, see the *Student Reference Book.*

Division Arrays

Players make arrays with counters. They use number cards to determine the number of counters and a toss of a die to establish the number of rows.

Beat the Calculator

A Calculator (a player who uses a calculator) and a Brain (a player who solves the problem without a calculator) compete to see who will be first to solve multiplication problems.

Do-Anytime Activities

To work with your child on concepts taught in this unit and in previous units,
try these interesting and rewarding activities:

1. Together with your child, sort objects into equal groups. Discuss what you could do
   with any leftover objects.

2. Review multiplication-fact shortcuts:

   ◆ **turn-around facts** The order of the factors does not change the product. Thus,
     if you know 3 × 4 = 12, you also know 4 × 3 = 12.

   ◆ **multiplication by 1** The product of 1 and another number is always equal to
     the other number. For example, 1 × 9 = 9; 1 × 7 = 7.

   ◆ **multiplication by 0** The product of 0 and another number is always 0. For
     example, 4 × 0 = 0; 0 × 2 = 0.

   ◆ **square numbers** Arrays for numbers multiplied by themselves are always
     squares. For example, 2 × 2 and 4 × 4 are square numbers.

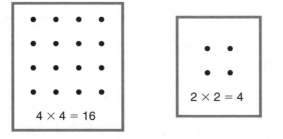

3. Use the ×, ÷ Fact Triangles (a set will be sent home later) to practice the basic facts.
   Act as a partner by covering one number on the card and then asking your child to
   create a multiplication or division number model using the other two numbers.

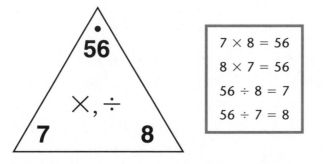

4. Write any number–for example, 34,056. Then ask questions like the following:
   How many are in the thousands place? *(4)* What is the value of the digit 5? *(50)*

5. Ask questions like the following:
   Is 467 + 518 more or less than 1,000? *(less)* Is 754 − 268 more or less than 500? *(less)*

As You Help Your Child with Homework

As your child brings home assignments, you might want to go over the instructions together, have your child explain the activities, and clarify them as necessary. The answers listed below will guide you through this unit's Home Links.

Home Link 4·1

1. 30 apples

Home Link 4·2

1. 24 counters **2.** 24 counters

3. 24 counters **4.** 358

5. 204 **6.** 428

Home Link 4·3

1. 5 counters per person; 0 counters remaining

2. 2 counters per person; 5 counters remaining

3. 4 weeks in January; 3 days remaining

4. 4 teams; 2 children remaining

5. 2 pencils; 4 pencils left over

6. 11 jelly beans; 0 jelly beans left over

7. 577 **8.** 31 **9.** 801

Home Link 4·4

1. 6 marbles; 0 marbles left over

2. 2 cookies; 1 cookie left over

3. 4 complete rows; 6 stamps left over

Home Link 4·5

1. 10; 10 **2.** 15; 15 **3.** 20; 20 **4.** 9; 9

5. 90; 90 **6.** 365; 365 **7.** 0; 0 **8.** 0; 0

9. 0; 0 **10.** 20 **11.** 20 **12.** 18

13. 14 **14.** 15 **15.** 50

Home Link 4·6

1. 10; 10; 10; 10 **2.** 12; 12; 12; 12

3. $2 \times 7 = 14$; $7 \times 2 = 14$; $14 \div 2 = 7$; $14 \div 7 = 2$

4. $2 \times 8 = 16$; $8 \times 2 = 16$; $16 \div 2 = 8$; $16 \div 8 = 2$

5. $5 \times 4 = 20$; $4 \times 5 = 20$; $20 \div 5 = 4$; $20 \div 4 = 5$

6. $10 \times 6 = 60$; $6 \times 10 = 60$; $60 \div 10 = 6$; $60 \div 6 = 10$

Home Link 4·7

1. $5 \times 6 = 30$; $6 \times 5 = 30$; $30 \div 6 = 5$; $30 \div 5 = 6$

2. $8 \times 3 = 24$; $3 \times 8 = 24$; $24 \div 3 = 8$; $24 \div 8 = 3$

3. $2 \times 9 = 18$; $9 \times 2 = 18$; $18 \div 2 = 9$; $18 \div 9 = 2$

4. $4 \times 7 = 28$; $7 \times 4 = 28$; $28 \div 7 = 4$; $28 \div 4 = 7$

5. $9 \times 8 = 72$; $8 \times 9 = 72$; $72 \div 9 = 8$; $72 \div 8 = 9$

6. $6 \times 7 = 42$; $7 \times 6 = 42$; $42 \div 7 = 6$; $42 \div 6 = 7$

Home Link 4·8

1. 7; 5; $7 \times 5 = 35$; 35 square units

2. 6; 7; $6 \times 7 = 42$; 42 square units

3. $4 \times 8 = 32$

4. $9 \times 5 = 45$

Home Link 4·9

The following answers should be circled:

1. more than the distance from Chicago to Dallas; about 2,400 miles

2. about 600 miles; less than the distance from Chicago to Denver

3. more than the distance from New York to Chicago

4. less than the distance from Denver to Atlanta; more than the distance from New York to Portland; about 750 miles

HOME LINK 4·1

Multiplication Number Stories

Family Note   Today your child learned about another tool to use when solving number stories, a multiplication/division diagram. It can help your child organize the information in a number story. With the information organized, your child can decide which operation (×, ÷) will solve the problem. Refer to pages 259 and 260 in the *Student Reference Book* for more information.

Please return this Home Link to school tomorrow.

SRB
259 260

For the number story:

◆ Fill in a multiplication/division diagram. Write ? for the number you will find. Then write the numbers you know.

◆ Use counters or draw pictures to help you find the answer.

◆ Write the answer and unit. Check whether your answer makes sense.

1. Elsa buys 5 packages of apples for the party. There are 6 apples in each package. How many apples does she have?

   Answer: _____
                     (unit)

   Does your answer make sense?

packages	apples per package	apples in all

2. Find equal groups of objects in your home, or around your neighborhood. Record them on the back of this page.
   Examples
   3 lights on each traffic light, 12 eggs per carton

3. Write a multiplication number story about one of your groups. Use the back of this paper. Solve the number story.

HOME LINK 4·2

Arrays

Family Note

Your child is learning how to represent multiplication problems using pictures called *arrays*. An array is a group of items arranged in equal rows and equal columns. Help your child use counters, such as pennies or macaroni, to build the array in each problem. Your child should record each solution on the dots next to the problem.

Please return this Home Link to school tomorrow.

SRB
64 65

For the next few weeks, look for pictures of items arranged in equal rows and columns, or **arrays.** Look in newspapers or magazines. Have people in your family help you. Explain that your class is making an Arrays Exhibit.

This is a 5-by-6 array. There are 5 rows. There are 6 dots in each row. There are 30 dots in all, since $5 \times 6 = 30$.

Make an array with counters. Mark the dots to show the array.

1. 4 rows with 6 counters per row

   a **4-by-6 array**

   _____ counters

2. 3 rows with 8 counters per row

   a **3 × 8 array**

   _____ counters

3. 2 rows with 12 counters per row

   a **2 × 12 array**

   _____ counters

Practice

Write these problems on the back of this page. Solve. Show your work.

4. $331 + 27 =$ _____

5. _____ $= 187 + 17$

6. $907 - 479 =$ _____

Unit

HOME LINK 4·3 Division with Counters

Family Note Your child is beginning to use division to solve number stories. A first step is to use counters, such as uncooked macaroni or pennies, to represent each problem. This helps your child to understand the meaning of division.

Please return this Home Link to school tomorrow.

SRB
73 74

Show someone at home how to do division using pennies, uncooked macaroni, or other counters.

1. 25 counters are shared equally by 5 people.

_____ counters per person

_____ counters remaining

2. 25 counters are shared equally by 10 people.

_____ counters per person

_____ counters remaining

3. 31 days in January
7 days per week

_____ weeks in January

_____ days remaining

4. 22 children
5 children per team

_____ teams

_____ children remaining

5. Mrs. Blair has 34 pencils to give to her 15 students. How many pencils can she give each student?

_____ pencils   _____ pencils left over

6. Caleb shared 22 jelly beans with his sister. How many jelly beans did each child get?

_____ jelly beans   _____ jelly beans left over

Practice

Write these problems on the back of this page. Solve. Show your work.

Unit

7. _____ = 614 − 37   **8.** 23 + 8 = _____

9. 123 + 678 = _____

83

HOME LINK 4·4

Division Number Stories

Family Note Help your child solve the division number stories by using counters such as pennies or uncooked macaroni to model the problems. Refer to pages 73, 74, 259, and 260 in the *Student Reference Book.* Your child is not expected to know division facts at this time.

Please return this Home Link to school tomorrow.

SRB
73 74
259 260

Use counters or draw pictures to show someone at home how you can use division to solve number stories. Fill in the diagrams.

1. Jamal gave 24 marbles to 4 friends. Each friend got the same number of marbles. How many marbles did each friend get?

friends	marbles per friend	marbles in all

_____ marbles

How many marbles were left over? _____ marble(s)

2. Ellie had 29 cookies to put in 14 lunch bags. She put the same number in each bag. How many cookies did she put in each bag?

bags	cookies per bag	cookies in all

_____ cookies

How many cookies were left over? _____ cookie(s)

3. A sheet of stamps has 46 stamps. A complete row has 10 stamps. How many complete rows are there?

complete rows	stamps per row	stamps in all

_____ complete rows

How many stamps were left over? _____ stamp(s)

Multiplication-Fact Shortcuts

> **Family Note**   Your child is learning the basic multiplication facts. Listen to your child explain multiplication-fact shortcuts as he or she works the problems. Review some 1s, 2s, 5s, and 10s multiplication facts (facts like $1 \times 3 = ?$, $? = 2 \times 4$, $5 \times 5 = ?$, and $10 \times 4 = ?$).
>
> *Please return this Home Link to school tomorrow.*

SRB
56

Tell someone at home about multiplication-fact shortcuts.

The turn-around rule: $3 \times 4 = 12$ helps me know $4 \times 3 = 12$.

1. $2 \times 5 =$ _____ and $5 \times 2 =$ _____

2. _____ $= 5 \times 3$ and _____ $= 3 \times 5$

3. $10 \times 2 =$ _____ and $2 \times 10 =$ _____

If 1 is multiplied by any number, the product is that number.
The same is true if any number is multiplied by 1.

4. _____ $= 1 \times 9$ and _____ $= 9 \times 1$

5. $1 \times 90 =$ _____ and $90 \times 1 =$ _____

6. $365 \times 1 =$ _____ and $1 \times 365 =$ _____

If 0 is multiplied by any number, the product is 0.
The same is true if any number is multiplied by 0.

7. $0 \times 12 =$ _____ and $12 \times 0 =$ _____

8. $99 \times 0 =$ _____ and $0 \times 99 =$ _____

9. _____ $= 9,365 \times 0$ and _____ $= 0 \times 9,365$

Think about counting by 2s, 5s, and 10s.

10.	**11.**	**12.**	**13.**	**14.**	**15.**
10	5	9	2	5	10
$\times 2$	$\times 4$	$\times 2$	$\times 7$	$\times 3$	$\times 5$

87

HOME LINK 4·6

×, ÷ **Fact Triangles**

Family Note

Fact Triangles build mental-math reflexes. They are the *Everyday Mathematics* version of traditional flash cards. Fact Triangles are better tools for memorizing, however, because they emphasize fact families.

A **fact family** is a group of facts made from the same 3 numbers. For 6, 4, and 24, the multiplication and division fact family is $4 \times 6 = 24$, $6 \times 4 = 24$, $24 \div 6 = 4$, $24 \div 4 = 6$.

Use Fact Triangles to practice basic facts with your child. Cut out the triangles from the two attached sheets.

To practice multiplication:

Cover the number under the large dot—the product.

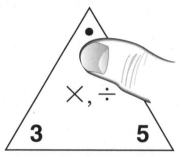

Your child should name one or two multiplication facts: $3 \times 5 = 15$, or $5 \times 3 = 15$.

To practice division, cover one of the smaller numbers.

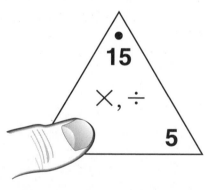

Your child should name the division fact $15 \div 5 = 3$.

Your child should name the division fact $15 \div 3 = 5$.

If your child misses a fact, flash the other two problems and then return to the fact that was missed. *Example:* Ravi can't answer $15 \div 3$. Flash 3×5, and then $15 \div 5$, and finally $15 \div 3$ a second time.

Make this activity brief and fun. Spend about 10 minutes each night for the next few weeks, or until your child learns them all. The work you do at home will support the work we are doing at school.

*Please return the **second page** of this Home Link to school tomorrow.*

SRB 54 55

HOME LINK
4·6 | ×, ÷ **Fact Triangles** *continued*

Tell someone at home about multiplication/division fact families.

1. The numbers 2, 5, and 10 form the following facts:

2 × 5 = _____   _____ ÷ 2 = 5

5 × 2 = _____   _____ ÷ 5 = 2

2. Knowing 6 × 2 = _____ and 2 × 6 = _____

helps me know _____ ÷ 2 = 6 and _____ ÷ 6 = 2.

3. The numbers 2, 7, and 14 form this multiplication/division fact family:

_____   _____

_____   _____

Write the fact family for each ×, ÷ Fact Triangle.

4.
16
×, ÷
2   8

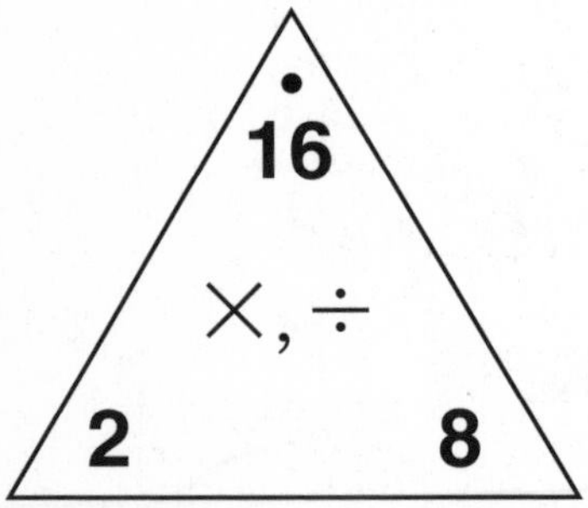

5.
20
×, ÷
5   4

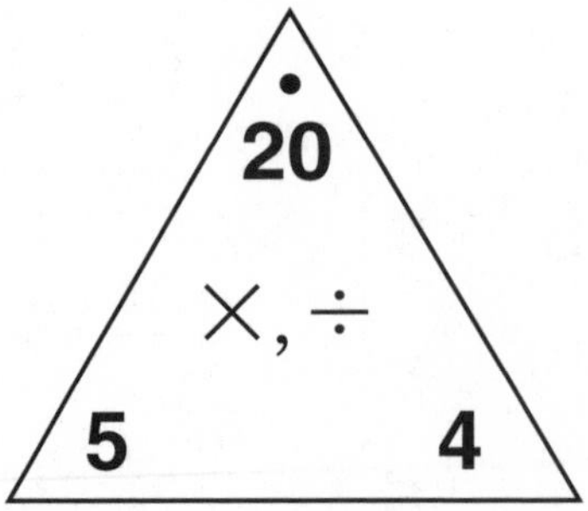

6.
60
×, ÷
10   6

_____ _____ _____

_____ _____ _____

_____ _____ _____

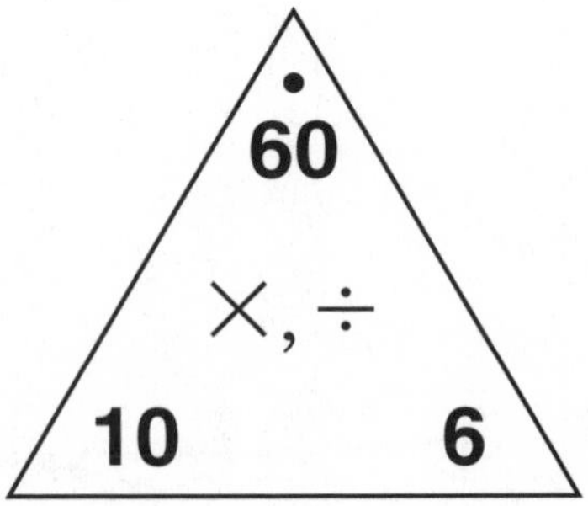

HOME LINK 4·7

Fact Families

Family Note   Your child continues to practice multiplication in school. You can help by stressing the relationship between multiplication and division: With the three nonzero numbers in a multiplication fact, two division facts can be formed. Fact Triangles are designed to help children understand this concept.

Please return this Home Link to school tomorrow.

SRB 54 55

Write the fact family for each Fact Triangle.

1.
30
×, ÷
5   6

____ × ____ = ____

____ × ____ = ____

____ ÷ ____ = ____

____ ÷ ____ = ____

2.
24
×, ÷
8   3

____ × ____ = ____

____ × ____ = ____

____ ÷ ____ = ____

____ ÷ ____ = ____

3.
18
×, ÷
2   9

____ × ____ = ____

____ × ____ = ____

____ ÷ ____ = ____

____ ÷ ____ = ____

4.
28
×, ÷
4   7

____ × ____ = ____

____ × ____ = ____

____ ÷ ____ = ____

____ ÷ ____ = ____

5.
72
×, ÷
9   8

____ × ____ = ____

____ × ____ = ____

____ ÷ ____ = ____

____ ÷ ____ = ____

6.
42
×, ÷
6   7

____ × ____ = ____

____ × ____ = ____

____ ÷ ____ = ____

____ ÷ ____ = ____

91

HOME LINK
4·8
Arrays and Areas

> **Family Note**   Your child uses the same procedure for finding the area of a rectangle that is used for finding the number of dots in an array. For Problem 3 it does not matter whether your child draws an array with 4 rows of 8 dots or 8 rows of 4 dots. What is important is that the array has two sides that have 4 dots and two sides that have 8 dots. The same concept is true for Problem 4.
>
> *Please return this Home Link to school tomorrow.*

SRB
64 65

Make a dot inside each small square in one row. Then fill in the blanks.

1. Number of rows: _____

Number of squares in a row: _____

Number model: _____ × _____ = _____

Area: _____ square units

2. Number of rows: _____

Number of squares in a row: _____

Number model: _____ × _____ = _____

Area: _____ square units

Mark the dots to show each array. Then fill in the blanks.

3. Make a 4-by-8 array.

Number model: _____ × _____ = _____

4. Make a 9-by-5 array.

Number model: _____ × _____ = _____

HOME LINK
4·9

Using a Map Scale

Family Note

Your child is just learning how to use a map scale. He or she should use the scale to measure an as-the-crow-flies estimate for each problem. This expression refers to the most direct route between two points, disregarding road distance. Actual road distances are longer than these direct paths.

Please return this Home Link to school tomorrow.

For each question, circle all reasonable answers. (There may be more than one reasonable answer.) All distances are as the crow flies. Be sure to use the map scale on the next page.

1. About how many miles is it from New York to Los Angeles?

    about 1,000 miles

    more than the distance from Chicago to Dallas

    about 2,400 miles

2. About how many miles is it from Chicago to Atlanta?

    about 600 miles

    more than the distance from Chicago to Seattle

    less than the distance from Chicago to Denver

3. About how many miles is it from Seattle to Dallas?

    about 2,600 miles

    about 5,000 miles

    more than the distance from New York to Chicago

4. About how many miles is it from New York to Atlanta?

    less than the distance from Denver to Atlanta

    more than the distance from New York to Portland

    about 750 miles

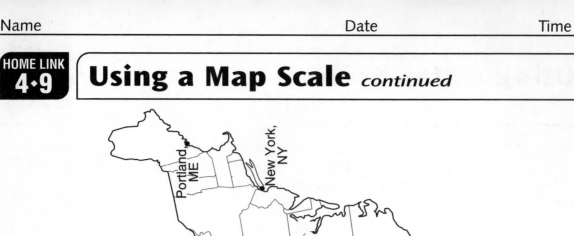

HOME LINK 4·9

Using a Map Scale *continued*

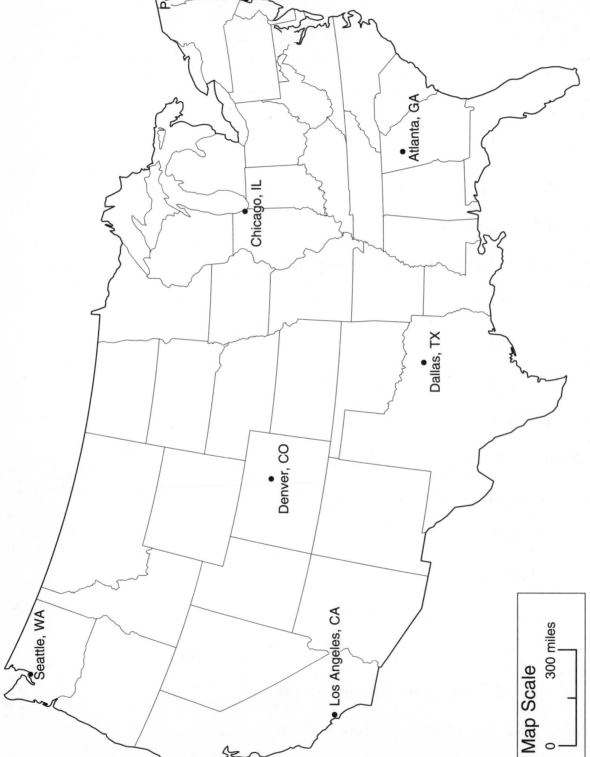

HOME LINK 4·10

A Fair Game?

Play the game *Rock, Paper, Scissors* with someone at home. Play at least 20 times. Keep a tally of wins and losses.

Rock, Paper, Scissors

Materials   ☐ players' hands

Players   2

Object of the Game   To choose a hand position that beats your partner's choice.

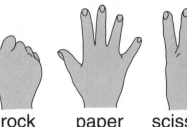

rock   paper   scissors

Directions

1. Each player hides one hand behind his or her back and puts it in the rock, paper, or scissors position.

2. One player counts, "One, two, three."

3. On "three," both players show their hand positions.

4. Players choose the winner according to these rules.

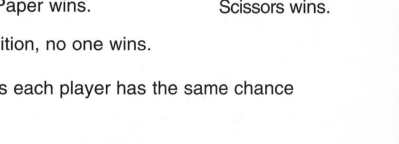

Rock dents scissors.   Paper covers rock.   Scissors cut paper.

Rock wins.   Paper wins.   Scissors wins.

If both players show the same position, no one wins.

1. Is this a fair game? (*Fair* means each player has the same chance

   of winning.) _____

2. On the back of this paper, explain why or why not.

97

HOME LINK
4·11

Unit 5: Family Letter

Place Value in Whole Numbers and Decimals

In Unit 5, children will review place value up to 5-digit whole numbers. They will read, write, compare, and order these numbers before they begin to explore larger numbers.

To understand real-life applications of larger numbers, children will study population data about U.S. cities. They will also approximate their own ages to the minute.

In second grade, children studied decimals by working with money. In this unit, they will gradually extend their knowledge of decimals in the following ways:

◆ through concrete models, such as base-10 blocks.

◆ by writing decimal values in three ways (0.1, one-tenth, $\frac{1}{10}$).

◆ by comparing and ordering numbers with symbols (<, >, =).

Decimal	Word	Fraction
0.1	one-tenth	$\frac{1}{10}$
0.2	two-tenths	$\frac{2}{10}$
0.3	three-tenths	$\frac{3}{10}$
0.4	four-tenths	$\frac{4}{10}$

Please keep this Family Letter for reference as your child works through Unit 5.

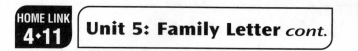

Vocabulary

Important terms in Unit 5:

place value A system that gives a digit a value according to its position, or place, in a number. The value of each digit in a numeral is determined by its place in the numeral. This chart demonstrates the value of each digit in the numeral 4,815.904 (read as *four thousand, eight hundred fifteen, and nine hundred four thousandths*):

thousands	hundreds	tens	ones		tenths	hundredths	thousandths
4	8	1	5	.	9	0	4
Each thousand is equal to one thousand times the unit value.	Each hundred is equal to one hundred times the unit value.	Each ten is equal to ten times the unit value.	Each one is equal to the unit value.		Each tenth is equal to $\frac{1}{10}$ of the unit value.	Each hundredth is equal to $\frac{1}{100}$ of the unit value.	Each thousandth is equal to $\frac{1}{1,000}$ of the unit value.
4,000	800	10	5		$\frac{9}{10}$	$\frac{0}{100}$	$\frac{4}{1,000}$

maximum The largest amount, or the greatest number in a set of data.

millimeter A metric unit of length equivalent to $\frac{1}{10}$ of a centimeter and $\frac{1}{1,000}$ of a meter.

pie graph A graph in which a circle is divided into regions corresponding to parts of a set of data.

Areas of the Continents
(in square miles)

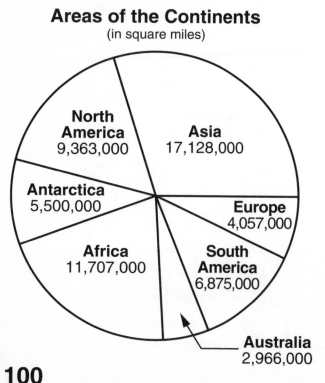

line graph A graph in which data points are connected by line segments.

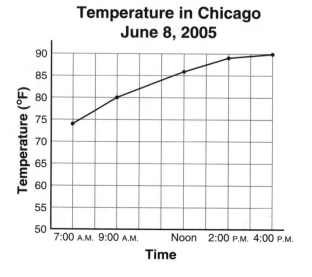

Temperature in Chicago June 8, 2005

Do-Anytime Activities

To work with your child on the concepts taught in this unit and in previous units, try these activities:

1. Dictate large numbers for your child to write. *Examples:* 4,123; 10,032; 2,368,502.

2. Display similar multidigit numbers on a calculator for your child to read.

3. Together, write 5 multidigit numbers in order from smallest to largest.

4. Start at any whole number and, using a calculator, count on by increments of 0.01 or 0.1.

5. Use money on a family shopping trip; practice making change.

Building Skills through Games

In Unit 5, your child will practice numeration and computation skills by playing the following games. For detailed instructions, see the *Student Reference Book.*

Baseball Multiplication

Players use multiplication facts to score runs. Team members take turns pitching by rolling two dice to get two factors. Then players on the batting team take turns multiplying the two factors and saying the product.

Number Top-It

As players pick each card, they must decide in which place-value box (from ones to ten-thousands at first, and then on to hundred-thousands) to place the card so that they end up with the largest number.

Beat the Calculator

A Calculator (a player who uses a calculator) and a Brain (a player who solves the problem without a calculator) race to see who will be first to solve multiplication problems.

Division Arrays

Players make arrays with counters using number cards to determine the number of counters and a toss of a die to determine the number of rows.

As You Help Your Child with Homework

As your child brings home assignments, you may want to go over the instructions together, clarifying them as necessary. The answers listed below will guide you through this unit's Home Links.

Home Link 5·1

1. 8,879; 8,889; 8,899; 8,909; 8,919; 8,929
2. 8,789; 8,889; 8,989; 9,089; 9,189; 9,289
3. 7,889; 8,889; 9,889; 10,889; 11,889; 12,889

Home Link 5·2

1. <   2. >   <
4. <   5. >   6. <
7. 3,689   8. 9,863   9. Answers vary.
10. 51,100; 52,100   11. 56
12. 163   13. 796   14. 484

Home Link 5·3

1. largest: 7,654,321; smallest: 1,234,567
   total: 8,888,888
3. 7,037,562; 7,000,007; 4,056,211; 104,719;
   42,876; 25,086; 9,603; 784
4. 42,876   5. 7,037,562
6. 4,056,211   7. 7,000,007

Home Link 5·4

1. 7 continents   2. Asia   3. Australia
4. Antarctica, North America, and South America
5. Europe   6. North America
7. Africa

Home Link 5·5

3,358	5,338
3,385	5,383
3,538	5,833
3,583	8,335
3,835	8,353
3,853	8,533

Home Link 5·7

1. $\frac{3}{10}$ or $\frac{30}{100}$; 0.3 or 0.30   2. $\frac{9}{100}$; 0.09
3. $\frac{65}{100}$; 0.65   4. 0.3; 0.65; 0.65
8. 0.04, 0.53, 0.8

Home Link 5·8

1. 57 hundredths; 5 tenths 7 hundredths
2. 70 hundredths; 7 tenths 0 hundredths
3. 4 hundredths; 0 tenths 4 hundredths
4. 0.23   5. 8.4   6. 30.20   7. 0.05
8. 0.4; 0.5; 0.6; 0.7; 0.8; 0.9
9. 0.04; 0.05; 0.06; 0.07; 0.08; 0.09
10. 503   11. 603

Home Link 5·9

1. 0.01; 0.02; 0.03; 0.04; 0.05; 0.06; 0.07; 0.08
2. 0.8; 0.9; 1.0; 1.1; 1.2; 1.3; 1.4

7. 27   8. 40   9. 0
10. 12   11. 9   12. 15

Home Link 5·10

1. a. 2 b. 10 c. 20 d. 100 e. 200 f. 600
2. a. 30 cm   b. 0.3 m   c. 300 mm
3. 49   4. 56   5. 63   6. 42

Home Link 5·11

1. <   2. <   3. >   4. =
5. >   6. <   7. =   8. <
9. hundredths, or 0.09   10. ones, or 3
11. 6.59; 6.60; 6.61   12. 1.03; 1.13; 1.23
13. 4.4   14. 4.17   15. 8.1   16. 5.53
17. 243   18. 782   19. 509

Home Link 5·12

1. 455   2. 455

HOME LINK 5·1 — Frames and Arrows

Family Note Have your child read and solve the three Frames-and-Arrows problems. Review the rule that is being used in each puzzle. Ask your child to look for patterns in the frames. For example, which digit changes when adding or subtracting 10? *(tens digit and hundreds digit change when moving from the 8,800s to the 8,900s)* 100? *(hundreds digit and thousands digit change when moving from the 8,000s to the 9,000s)* 1,000? *(thousands digit and ten-thousands digit change when moving from the 9,000s to the 10,000s)*

Please return this Home Link to school tomorrow.

Solve each Frames-and-Arrows problem.

1.

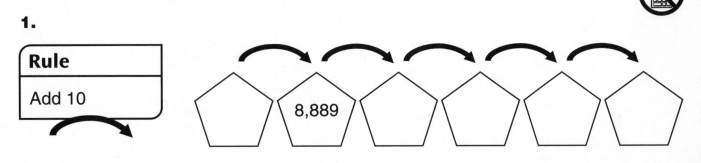

Rule
Add 10

8,889

2.

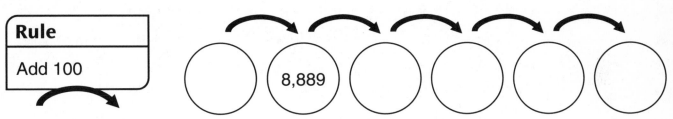

Rule
Add 100

8,889

3.

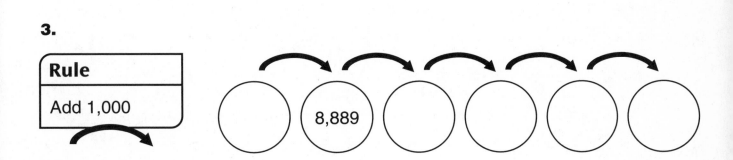

Rule
Add 1,000

8,889

HOME LINK 5·2 Comparing Numbers

Family Note

Review the meanings of the > and < relation symbols (see box below) before your child begins this page. When your child has completed the Home Link, ask him or her to read the numbers on the page to you.

The game *Number Top-It* gives children the opportunity to practice comparing 5-digit numbers. You may wish to play *Number Top-It* with your child. (See *Student Reference Book,* pages 302 and 303.)

Please return this Home Link to school tomorrow.

SRB
302 303

Write < or >.

1. 906 _____ 960

2. 5,708 _____ 599

3. 31,859 _____ 31,958

4. 10,006 _____ 10,106

5. 48,936 _____ 4,971

6. 76,094 _____ 76,111

> < means *is less than*
>
> > means *is greater than*

Use the digits 6, 8, 3, and 9.

7. Write the smallest possible number. _____

8. Write the largest possible number. _____

9. Write two numbers that are between the smallest and largest numbers.

_____   _____

10. Fill in the missing numbers.

50,100   _____   _____   53,100

Practice

Write these problems on the back of this page. Solve. Show your work.

11. 48
　　+ 8

12. 86
　　+77

13. 717
　　+ 79

14. 236
　　+248

HOME LINK 5·3

Practice with Place Value

Family Note Help your child use the seven digit squares to make the largest and smallest whole numbers possible out of all seven digits. *Number Top-It* (7-Digit Numbers) on *Student Reference Book*, page 304 provides practice comparing 7-digit numbers. You may wish to play this game with your child.

Please return this Home Link to school tomorrow.

SRB
304

1. Cut out the digit squares. Use all 7 digits to make the largest number and the smallest. Add the numbers and then read them to someone at home.

   largest _____

   smallest _____

   Total _____

2. Read the following numbers to someone at home:

   784   25,086   4,056,211   42,876

   9,603   7,000,007   7,037,562   104,719

3. Write the numbers above in order from the largest to the smallest.

   _____
   (largest)

   _____

   _____

   _____

   _____

   _____
   (smallest)

4. Which number is 1,000 less than 43,876?

   _____

5. Which number is 10,000 more than 7,027,562?

   _____

6. Which number is 10,000 less than 4,066,211?

   _____

7. Which number is 1,000,000 more than 6,000,007?

   _____

6

2

4

7

1

5

3

HOME LINK 5·4

Comparing Areas of Continents

Family Note   Your child has been practicing reading and writing 6- and 7-digit numerals. Use the pie graph to help him or her answer the questions about the continents. Ask your child to read each of the areas of the continents aloud to you. Encourage rounding the areas to the nearest million when making the comparisons in Problems 5–7. Remember that working with numbers in the millions is a new skill for your child.

Please return this Home Link to school tomorrow.

SRB
194

Use the graph to answer the questions.

Areas of the Continents
(in square miles)

1. How many continents are there?

_____

2. Which continent has the largest area?

_____

3. Which continent has the smallest area?

_____

4. Which continents have an area between 5 and 10 million square miles each?

_____

_____

North America 9,363,000

Asia 17,128,000

Antarctica 5,500,000

Europe 4,057,000

Africa 11,707,000

South America 6,875,000

Australia 2,966,000

5. Which continent is about 1 million square miles larger than Australia?

_____

6. Which continent is a little more than half the size of Asia?

_____

Try This

7. Which continent is a little less than 3 times the size of Europe?

_____

109

HOME LINK 5·5

Writing and Ordering Numbers

Family Note   Observe and encourage as your child makes 4-digit numbers using the digit squares, records the numbers, and then writes them in order from smallest to largest. Then listen as your child reads the numbers to you.

Please return this Home Link to school tomorrow.

Cut out the digit squares. Arrange them into 4-digit numbers in as many different ways as you can. Record each number you make. Then put the numbers in order from smallest to largest. Read your numbers to someone at home.

Record numbers here: **Order** numbers here:

_____

(smallest)

_____   _____

_____   _____

_____   _____

_____   _____

_____   _____

_____   _____

_____   _____

_____   _____

_____

(largest)

| 3 | 5 | 8 | 3 |

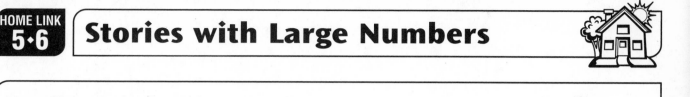

Stories with Large Numbers

Family Note Help your child write an addition and a subtraction story using 5-, 6-, or 7-digit numbers. Your child has been working with numbers as large as millions (7 digits), so this is a realistic expectation. However, it is acceptable for children to make up stories with 5- or 6-digit numbers.

Please return this Home Link to school tomorrow.

For each number story, try to think about large numbers of things. Share your stories with someone at home. If the numbers are too big for you to add or subtract, use a calculator or ask someone at home to help.

1. Write a number story that you solve by adding.   **Workspace**

_____

_____

_____

_____

Answer: _____
                    (unit)

2. Write a number story that you solve by subtracting.

_____

_____

_____

_____

Answer: _____
                    (unit)

113

Understanding Decimals

Family Note Your child has been using grids like the ones below to understand the meaning of decimals. The grid is made up of 100 squares. Each square is $\frac{1}{100}$ or 0.01 of the grid. Ten squares is $\frac{1}{10}$ or 0.10 of the grid. 0.8 is read as "eight-tenths." 0.04 is read as "four-hundredths." 0.53 is read as "fifty-three hundredths."

Please return this Home Link to school tomorrow.

SRB
33–36

If the grid is ONE, then what part of each grid is shaded?
Write a fraction and a decimal below each grid.

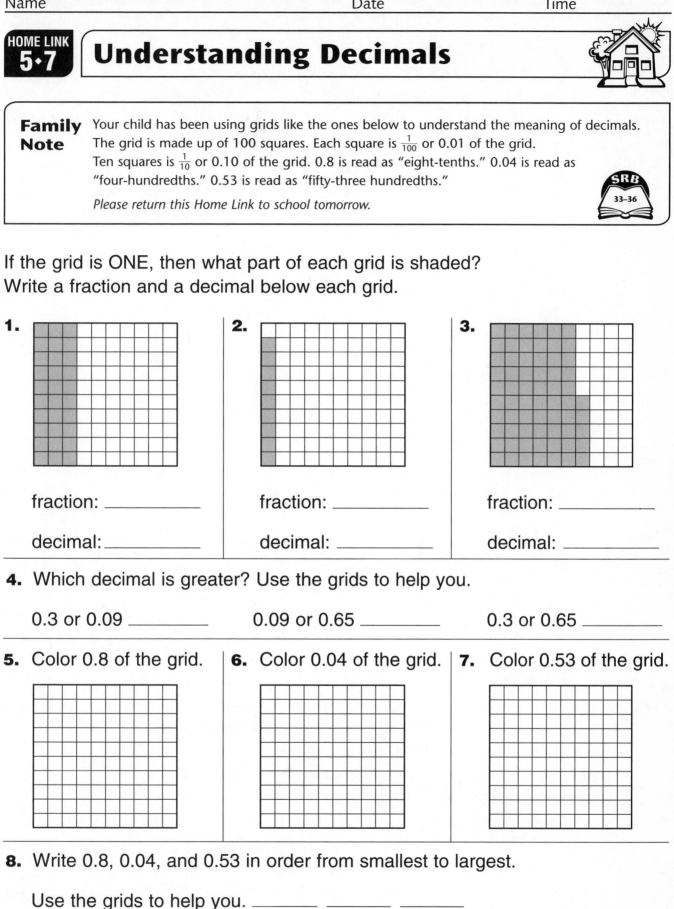

1.

fraction: _____

decimal: _____

2.

fraction: _____

decimal: _____

3.

fraction: _____

decimal: _____

4. Which decimal is greater? Use the grids to help you.

0.3 or 0.09 _____   0.09 or 0.65 _____   0.3 or 0.65 _____

5. Color 0.8 of the grid.

6. Color 0.04 of the grid.

7. Color 0.53 of the grid.

8. Write 0.8, 0.04, and 0.53 in order from smallest to largest.

Use the grids to help you. _____ _____ _____

HOME LINK 5·8 Tenths and Hundredths

Family Note Your child continues to work with decimals. Encourage him or her to think about ways to write money amounts. This is called dollars-and-cents notation. For example, $0.07 (7 cents), $0.09 (9 cents), and so on.

Please return this Home Link to school tomorrow.

SRB
33–35

Write what each diagram shows.

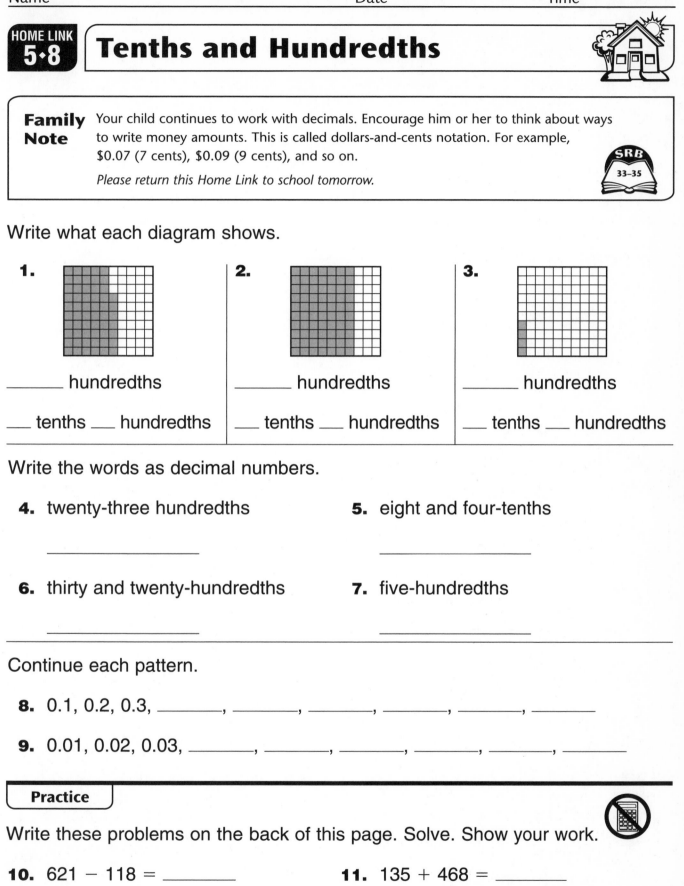

1.

_____ hundredths

___ tenths ___ hundredths

2.

_____ hundredths

___ tenths ___ hundredths

3.

_____ hundredths

___ tenths ___ hundredths

Write the words as decimal numbers.

4. twenty-three hundredths

_____

5. eight and four-tenths

_____

6. thirty and twenty-hundredths

_____

7. five-hundredths

_____

Continue each pattern.

8. 0.1, 0.2, 0.3, _____, _____, _____, _____, _____, _____

9. 0.01, 0.02, 0.03, _____, _____, _____, _____, _____, _____

Practice

Write these problems on the back of this page. Solve. Show your work.

10. 621 − 118 = _____

11. 135 + 468 = _____

HOME LINK	
5·9	**Practice with Decimals**

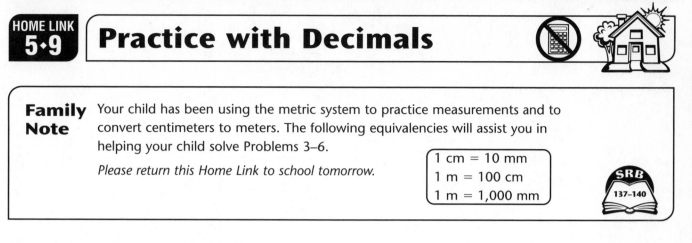

Family Note Your child has been using the metric system to practice measurements and to convert centimeters to meters. The following equivalencies will assist you in helping your child solve Problems 3–6.

Please return this Home Link to school tomorrow.

1 cm = 10 mm
1 m = 100 cm
1 m = 1,000 mm

SRB
137–140

Fill in the missing numbers.

1.

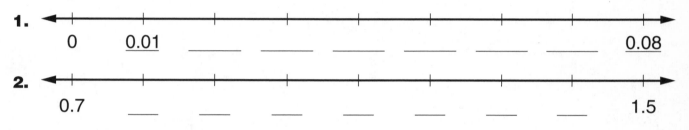

0   0.01   ____   ____   ____   ____   ____   ____   0.08

2.

0.7   ___   ___   ___   ___   ___   ___   ___   1.5

Follow these directions on the ruler below.

3. Make a dot at 7 cm and label it with the letter *A*.

4. Make a dot at 90 mm and label it with the letter *B*.

5. Make a dot at 0.13 m and label it with the letter *C*.

6. Make a dot at 0.06 m and label it with the letter *D*.

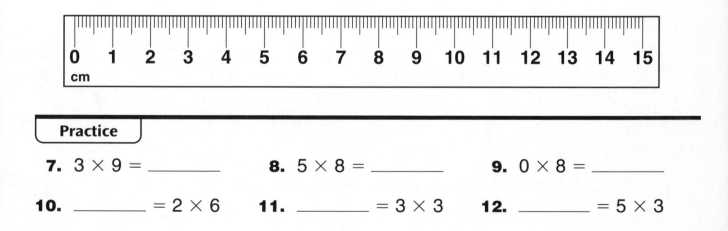

Practice	

7. $3 \times 9 =$ _____ **8.** $5 \times 8 =$ _____ **9.** $0 \times 8 =$ _____

10. _____ $= 2 \times 6$ **11.** _____ $= 3 \times 3$ **12.** _____ $= 5 \times 3$

HOME LINK 5·10

Measuring with Millimeters

Family Note Your child has been using millimeters to learn about decimal place value. This page offers a way to practice with millimeters and other metric measurements. Have your child use the ruler at the bottom of the page to answer the questions.

Please return this Home Link to school tomorrow.

SRB
137–140

A queen termite is drawn above the ruler at the bottom of the page. It is 5 millimeters long.

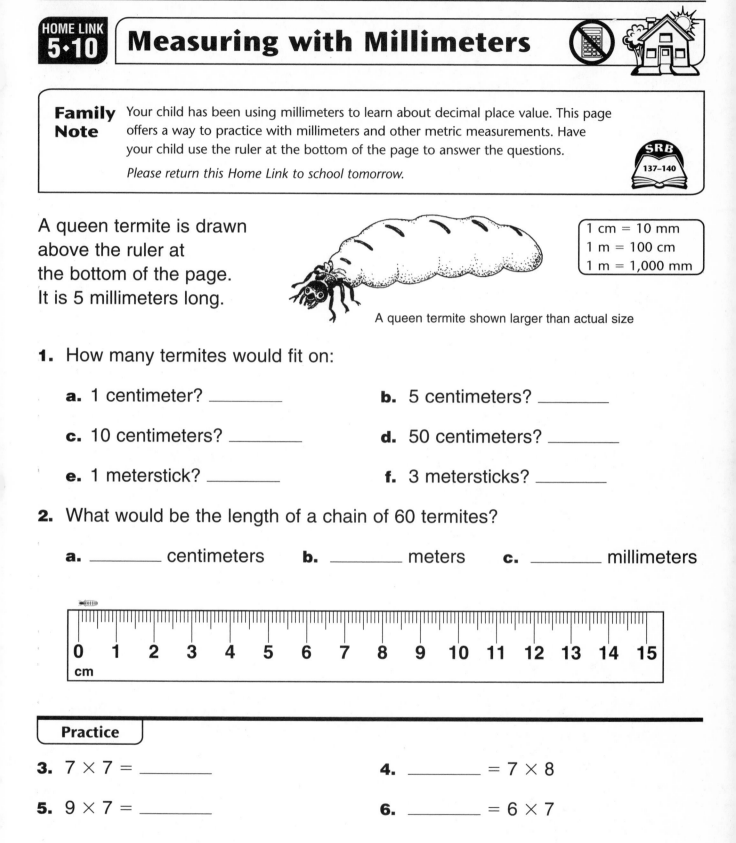

A queen termite shown larger than actual size

| 1 cm = 10 mm |
| 1 m = 100 cm |
| 1 m = 1,000 mm |

1. How many termites would fit on:

   a. 1 centimeter? _____   **b.** 5 centimeters? _____

   c. 10 centimeters? _____   **d.** 50 centimeters? _____

   e. 1 meterstick? _____   **f.** 3 metersticks? _____

2. What would be the length of a chain of 60 termites?

   a. _____ centimeters   **b.** _____ meters   **c.** _____ millimeters

0 1 2 3 4 5 6 7 8 9 10 11 12 13 14 15
cm

Practice

3. $7 \times 7 =$ _____   4. _____ $= 7 \times 8$

5. $9 \times 7 =$ _____   6. _____ $= 6 \times 7$

HOME LINK 5·11 Comparing Decimals

Family Note Ask your child to read the decimal numerals aloud. Encourage your child to use the following method:
1. Read the whole-number part.
2. Say *and* for the decimal point.
3. Read the digits after the decimal point as though they formed their own number.
4. Say *tenths, hundredths,* or *thousandths,* depending on the placement of the right-hand digit. Encourage your child to exaggerate the *ths* sound.

Please return this Home Link to school tomorrow.

SRB 35 36

Write >, <, or =.

> means *is greater than*

< means *is less than*

1. 2.35 _____ 2.57

2. 1.008 _____ 1.8

3. 0.64 _____ 0.46

4. 0.90 _____ 0.9

5. 42.1 _____ 42.09

6. 7.098 _____ 7.542

7. 0.4 _____ 0.400

8. 0.206 _____ 0.214

Example: The 4 in 0.47 stands for 4 __tenths__ or __0.4__.

9. The 9 in 4.59 stands for 9 _____ or _____.

10. The 3 in 3.62 stands for 3 _____ or _____.

Continue each number pattern.

11. 6.56, 6.57, 6.58, _____, _____, _____

12. 0.73, 0.83, 0.93, _____, _____, _____

Write the number that is 0.1 more.   Write the number that is 0.1 less.

13. 4.3 _____   **14.** 4.07 _____   **15.** 8.2 _____   **16.** 5.63 _____

Practice

Solve these problems on the back of this page. Show your work.

17. 282
      − 39

18. 811
      − 29

19. 685
      − 176

123

HOME LINK 5·12 | Subtraction & Multiplication Practice

Family Note Ask your child to explain the counting-up and trade-first subtraction methods.
Please return this Home Link to school tomorrow.

Make a ballpark estimate. Subtract and show your work. Check to see if your answer makes sense.

1. Use the counting-up method. _____
(Ballpark estimate)

```
  754
 −299
```

Unit

2. Use the trade-first method. _____
(Ballpark estimate)

```
  754
 −299
```

Multiplication. Write facts that you know.

3. × 2 facts

$4 \times 2 = 8$

_____

_____

_____

_____

4. × 3 facts

_____

_____

_____

_____

_____

5. × 4 facts

_____

_____

_____

_____

_____

125

HOME LINK 5·13

Unit 6: Family Letter

Geometry

Everyday Mathematics uses children's experiences with the everyday world to help them envision 3-dimensional (3-D) shapes. In previous grades, children were asked to identify 2-dimensional (2-D) shapes and their parts, such as sides and corners (vertices). They had several hands-on experiences with pattern blocks, geoboards, and templates. They also classified and named polygons, or closed figures consisting of line segments (sides) connected endpoint to endpoint.

In Unit 6, children will explore points, line segments, rays, lines, and the relationships among them, along with the geometric shapes that can be built from them. Children will construct angles, polygons, prisms, and pyramids.

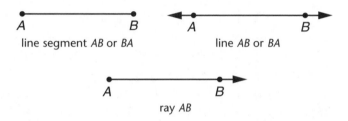

Children will also explore similarities and differences among 3-D shapes and regular polyhedrons within the context of a Shapes Museum. They will discover real-life examples of lines that are parallel, or lines that never meet, such as railroad tracks.

There is a great deal of vocabulary involved when working with geometry. However, the emphasis in this unit is not on memorizing the vocabulary, but rather on using it to examine relationships among classifications of geometric figures.

Please keep this Family Letter for reference as your child works through Unit 6.

Vocabulary

Important terms in Unit 6:

2-dimensional (2-D) shape A shape whose points are all in one plane, or flat surface, but not all on one line. A shape with length and width, but no thickness.

3-dimensional (3-D) shape A shape that does not lie completely within a plane, or flat surface; a shape with length, width, and thickness.

base of a 3-D shape A flat surface or face whose shape is the basis for naming some 3-dimensional objects.

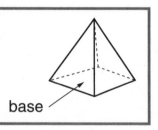

base

cone A 3-dimensional shape with a circular base, a curved surface, and one vertex, called the apex. An ice-cream cone is shaped like a cone.

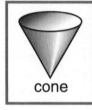

cone

sphere A 3-dimensional shape whose curved surface is, at all points, a given distance from its center point. A ball is shaped like a sphere.

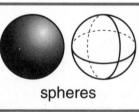

spheres

cylinder A 3-dimensional shape with two circular bases that are parallel and congruent and are connected by a curved surface. A soup can is shaped like a cylinder.

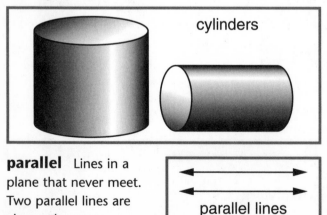

cylinders

parallel Lines in a plane that never meet. Two parallel lines are always the same distance apart.

parallel lines

face In *Everyday Mathematics,* a flat surface on a 3-dimensional shape.

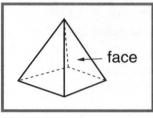

face

polyhedron

A *3-dimensional shape* with polygons and their interiors for *faces.* Polyhedrons don't have any holes. Below are five regular polyhedrons, so called because all faces in each shape are identical.

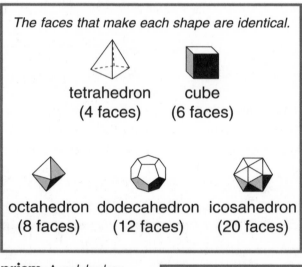

The faces that make each shape are identical.

tetrahedron (4 faces)   cube (6 faces)

octahedron (8 faces)   dodecahedron (12 faces)   icosahedron (20 faces)

prism A *polyhedron* with two parallel *bases* that are the same size and shape. A prism is named for the shape of its base, and the other faces are all parallelograms.

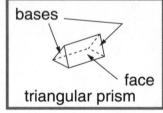

bases

face

triangular prism

pyramid A *polyhedron* with a polygon for a *base* and the other faces are all triangles with a common vertex called the apex. A pyramid is named for the shape of its base.

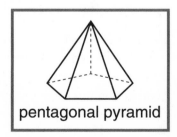

pentagonal pyramid

128

Do-Anytime Activities

To work with your child on the concepts taught in this unit and in previous units, try these interesting and rewarding activities:

1. Together, read the book *The Greedy Triangle,* by Marilyn Burns.

2. Begin a Shapes Museum at home. Label the shapes that your child collects.

3. Ask your child to identify 2-dimensional and 3-dimensional shapes inside and outside your home.

4. Measure objects to the nearest $\frac{1}{2}$ inch.

Building Skills through Games

In Unit 6, your child will practice numeration, multiplication, and geometry skills by playing the following games. For detailed instructions, see the *Student Reference Book.*

Number Top-It (Decimals)

As players pick each card, they must decide in which place-value box (from ones to thousandths) to place the card so that they end up with the largest number.

Beat the Calculator

A "Calculator" (a player who uses a calculator to solve the problem) and a "Brain" (a player who solves the problem without a calculator) race to see who will be first to solve multiplication problems.

Baseball Multiplication

Players use multiplication facts to score runs. Team members take turns "pitching" by rolling two dice to get two factors. Then players on the "batting" team take turns multiplying the two factors and saying the product.

Angle Race

Players build angles with rubber bands and "race" to see who will be first to complete the last angle exactly on the 360° mark.

As You Help Your Child with Homework

As your child brings home assignments, you may want to go over the instructions together, clarifying them as necessary. The answers listed below will guide you through this unit's Home Links.

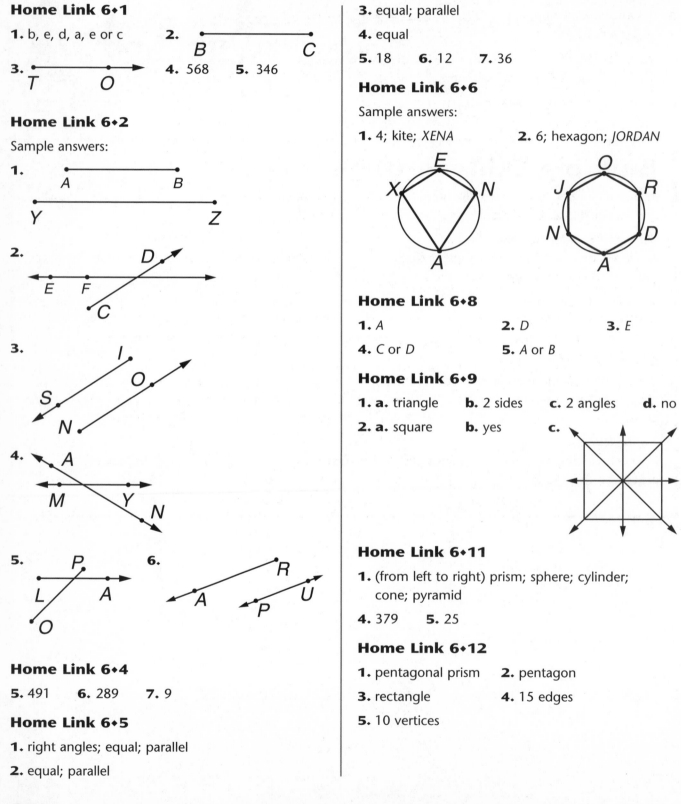

Home Link 6·1

1. b, e, d, a, e or c

2. *B* — — — *C*

3. *T* — — *O*

4. 568   **5.** 346

Home Link 6·2

Sample answers:

1. *A* — — *B*, *Y* — — *Z*

2. *D*, *E*, *F*, *C*

3. *I*, *O*, *S*, *N*

4. *A*, *M*, *Y*, *N*

5. *P*, *L*, *A*, *O*

6. *R*, *A*, *P*, *U*

Home Link 6·4

5. 491   **6.** 289   **7.** 9

Home Link 6·5

1. right angles; equal; parallel

2. equal; parallel

3. equal; parallel

4. equal

5. 18   **6.** 12   **7.** 36

Home Link 6·6

Sample answers:

1. 4; kite; *XENA*   **2.** 6; hexagon; *JORDAN*

Home Link 6·8

1. *A*   **2.** *D*   **3.** *E*

4. *C* or *D*   **5.** *A* or *B*

Home Link 6·9

1. a. triangle   **b.** 2 sides   **c.** 2 angles   **d.** no

2. a. square   **b.** yes   **c.**

Home Link 6·11

1. (from left to right) prism; sphere; cylinder; cone; pyramid

4. 379   **5.** 25

Home Link 6·12

1. pentagonal prism   **2.** pentagon

3. rectangle   **4.** 15 edges

5. 10 vertices

HOME LINK
6·1

Line Segments, Rays, and Lines

Family Note Help your child match each name below with the correct drawing of a line, ray, or line segment. Then observe as your child uses a straightedge to draw and label figures. Pages 100 and 101 in the *Student Reference Book* discuss these figures.

Please return this Home Link to school tomorrow.

SRB
100 101

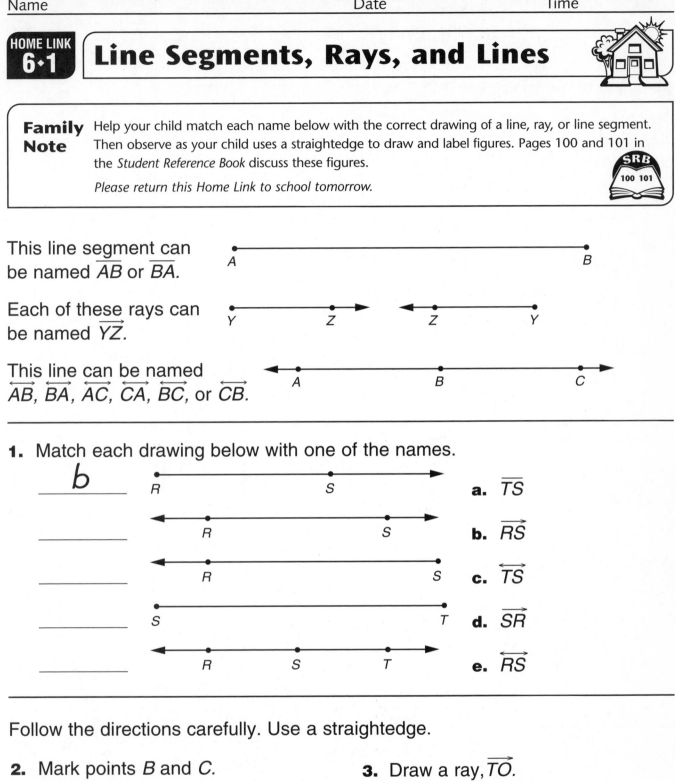

This line segment can be named $\overline{AB}$ or $\overline{BA}$.

Each of these rays can be named $\overrightarrow{YZ}$.

This line can be named $\overleftrightarrow{AB}$, $\overleftrightarrow{BA}$, $\overleftrightarrow{AC}$, $\overleftrightarrow{CA}$, $\overleftrightarrow{BC}$, or $\overleftrightarrow{CB}$.

1. Match each drawing below with one of the names.

   _____ **b**   R ●————————● S ————————►   **a.** $\overline{TS}$

   _____   ◄————● R ————————● S ————►   **b.** $\overrightarrow{RS}$

   _____   ◄————● R ————————● S   **c.** $\overleftrightarrow{TS}$

   _____   S ●————————————————● T   **d.** $\overrightarrow{SR}$

   _____   ◄————● R ————● S ————● T ————►   **e.** $\overleftrightarrow{RS}$

Follow the directions carefully. Use a straightedge.

2. Mark points *B* and *C*.
   Draw a line segment, $\overline{BC}$.

3. Draw a ray, $\overrightarrow{TO}$.

Practice

Write these problems on the back of this page. Solve.

4. 479 + 89 = _____

5. 278 + 68 = _____

HOME LINK 6·2 — More Line Segments, Rays, and Lines

Family Note Refer to the following notations to help your child draw and label line segments, rays, and lines.

line segment AB	$\overline{AB}$	
ray BA	$\overrightarrow{BA}$	
line AB	$\overleftrightarrow{AB}$	

Please return this Home Link to school tomorrow.

SRB
99–101

Use a straightedge and a sharp pencil to draw the following. Be sure to mark points and label the line segments, rays, and lines.

1. Draw line segment $\overline{YZ}$, that is parallel to $\overline{AB}$.

A ———————— B

2. Draw a ray, $\overrightarrow{CD}$, that intersects $\overleftrightarrow{EF}$.

E   F

3. Draw two parallel rays, $\overrightarrow{IS}$ and $\overrightarrow{NO}$.

4. Draw two intersecting lines, $\overleftrightarrow{MY}$ and $\overleftrightarrow{AN}$.

5. Draw a line segment $\overline{PO}$ intersecting ray $\overrightarrow{LA}$.

6. Draw line $\overleftrightarrow{PU}$, parallel to ray $\overrightarrow{RA}$.

133

HOME LINK 6·3

Right Angles

Family Note   Our class has been studying intersecting lines including lines that intersect at right angles. Help your child look for objects that have square corners or right angles—tables, pictures, the kitchen counter, a book, and so on.

Please return this Home Link to school tomorrow.

Find 4 things at home that have right angles (square corners).

Below, describe or draw a picture of each of these things. Bring your descriptions or your pictures to school to add to your Geometry Hunt.

HOME LINK
6·4

Triangles

Family Note Your child has been learning about the properties of triangles. Watch as your child completes the page.

SRB
106 107

For each problem, use a straightedge to connect the three points with three line segments. Show someone at home that the triangles match their descriptions. To measure triangles 1–3, cut out and use the ruler at the right. To find the right angle in triangle 4, use the square corner of a piece of paper.

1. equilateral triangle

All sides and angles are equal.

A•

•   •
B   C

2. isosceles triangle

Two sides are equal.

D•

•   •
F   E

3. scalene triangle

No sides are equal.

• G

I•

•H

4. right triangle

The triangle has a right angle ($\frac{1}{4}$ turn).

• K

J •

• L

15 14 13 12 11 10 9 8 7 6 5 4 3 2 1 0   CENTIMETERS

Practice

Solve the following problems on the back of this page.

5. $584 - 93 =$ _____   **6.** $823 - 534 =$ _____   **7.** _____ $= 234 - 225$

137

HOME LINK 6·5

Quadrangles

Family Note   Help your child complete the statements. A *right angle* is a square corner. *Parallel sides* are the same distance apart and will never meet. *Opposite sides* are directly across from each other. *Adjacent sides* meet at a vertex (corner).

Please return this Home Link to school tomorrow.

SRB 108 109

Fill in the blanks using the following terms: **equal**   **parallel**   **right angles**

1. Rectangle (Squares are special rectangles.)

All angles are _____.

Pairs of opposite sides are _____ in

length and _____ to each other.

2. Rhombus (Squares are also rhombuses.)

All sides are _____ in length.

Opposite sides are _____ to each other.

3. Parallelogram (Squares and rhombuses are also parallelograms.)

Opposite sides are _____ in length.

Opposite sides are _____ to each other.

4. Kite

Opposite sides are not _____ in length.

Practice

Solve.

5. $6 \times 3 =$ _____   **6.** _____ $= 3 \times 4$ **7.** $6 \times 6 =$ _____

139

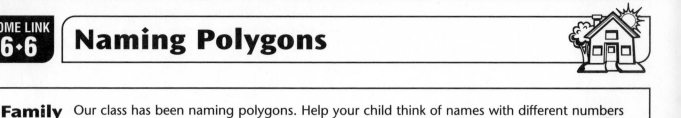

HOME LINK 6·6 Naming Polygons

Family Note Our class has been naming polygons. Help your child think of names with different numbers of letters, so that he or she can draw and name several different polygons.

Please return this Home Link to school tomorrow.

SRB
102 103

Think of names that have *different* letters. Use the letters to name points on each circle. Then use a pencil and a straightedge to connect the points to make a polygon. Count the number of sides. Name the polygon.

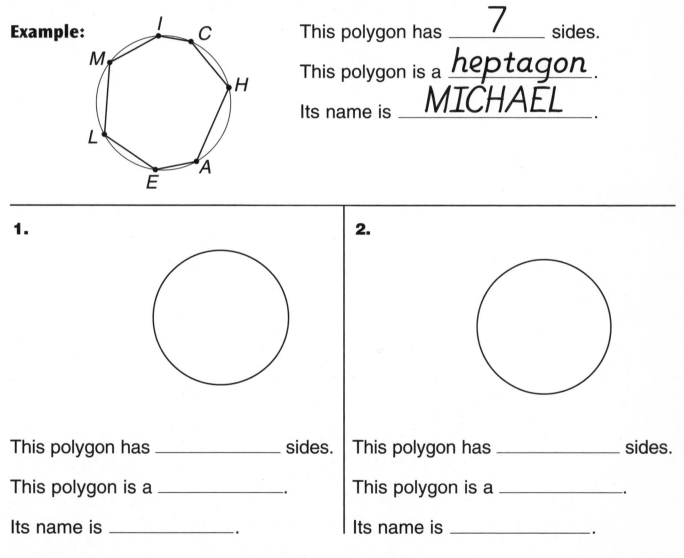

Example:

This polygon has _____7_____ sides.

This polygon is a _*heptagon*_.

Its name is _MICHAEL_.

1.

This polygon has _____ sides.

This polygon is a _____.

Its name is _____.

2.

This polygon has _____ sides.

This polygon is a _____.

Its name is _____.

3. Draw more circles and polygons on the back of this paper. Why do you think each letter in a polygon's name can be used only once?

_____

_____

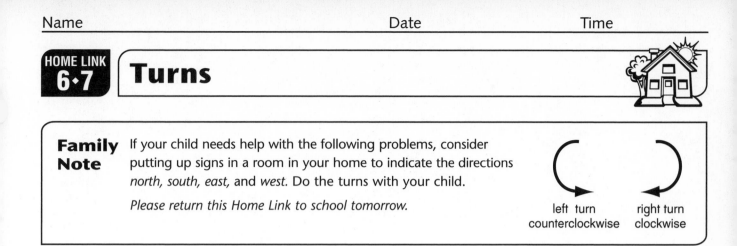

HOME LINK 6·7

Turns

Make the turns described below. Show which way you face after each turn.

◆ Draw a dot on the circle.

◆ Label the dot with a letter.

Example: Face north.

Do a $\frac{1}{2}$ turn counterclockwise.

On the circle, mark the direction you are facing with the letter *A*.

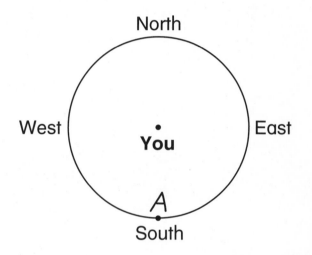

1. Face north. Do a $\frac{1}{4}$ turn clockwise. Mark the direction you are facing with the letter *B*.

2. Face north. Do a $\frac{3}{4}$ turn clockwise. Mark the direction you are facing with the letter *C*.

3. Face east. Do a $\frac{1}{4}$ turn counterclockwise. Mark the direction you are facing with the letter *D*.

4. Face west. Make less than a $\frac{1}{4}$ turn clockwise. Mark the direction you are facing with the letter *E*.

5. Face north. Make a clockwise turn that is more than a $\frac{1}{2}$ turn, but less than a $\frac{3}{4}$ turn. Mark the direction you are facing with the letter *F*.

6. Face north. Make a counterclockwise turn that is less than a $\frac{1}{2}$ turn, but more than a $\frac{1}{4}$ turn. Mark the direction you are facing with the letter *G*.

143

HOME LINK 6·8

Degree Measures

Family Note Our class has been learning about turns, angles, and angle measures. A full turn can be represented by an angle of 360°, a $\frac{1}{2}$ turn by an angle of 180°, a $\frac{1}{4}$ turn by an angle of 90°, and so on. Help your child match the measures below with the angles pictured. (It is not necessary to measure the angles with a protractor.)

Please return this Home Link to school tomorrow.

Tell which angle has the given measure.

1. about 180°   angle _____

2. about 90°   angle _____

3. about 270°   angle _____

4. between 0° and 90° angle _____

5. between 90° and 180° angle _____

Rotation	Degrees
$\frac{1}{4}$ turn	90°
$\frac{1}{2}$ turn	180°
$\frac{3}{4}$ turn	270°
full turn	360°

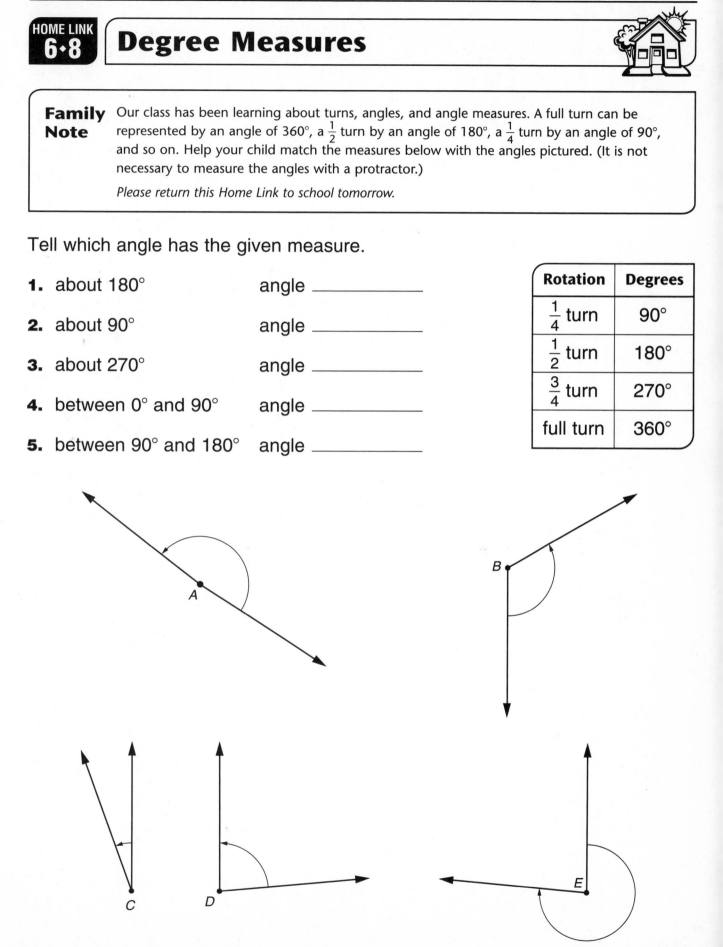

HOME LINK
6·9

Symmetric Shapes

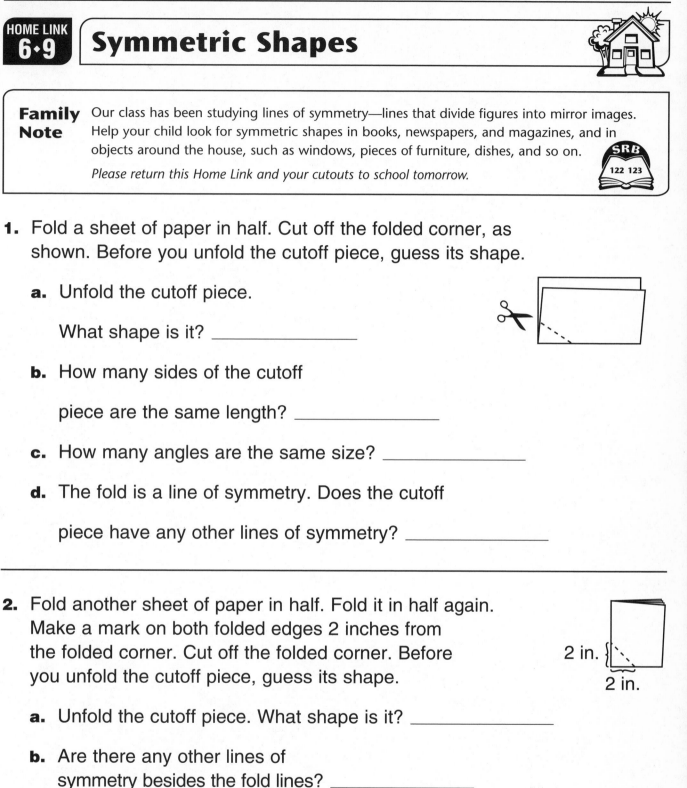

Family Note   Our class has been studying lines of symmetry—lines that divide figures into mirror images. Help your child look for symmetric shapes in books, newspapers, and magazines, and in objects around the house, such as windows, pieces of furniture, dishes, and so on.

Please return this Home Link and your cutouts to school tomorrow.

SRB
122 123

1. Fold a sheet of paper in half. Cut off the folded corner, as shown. Before you unfold the cutoff piece, guess its shape.

   a. Unfold the cutoff piece.

      What shape is it? _____

   b. How many sides of the cutoff

      piece are the same length? _____

   c. How many angles are the same size? _____

   d. The fold is a line of symmetry. Does the cutoff

      piece have any other lines of symmetry? _____

2. Fold another sheet of paper in half. Fold it in half again. Make a mark on both folded edges 2 inches from the folded corner. Cut off the folded corner. Before you unfold the cutoff piece, guess its shape.

   2 in.
   2 in.

   a. Unfold the cutoff piece. What shape is it? _____

   b. Are there any other lines of symmetry besides the fold lines? _____

   c. On the back of this paper, draw a picture of the cutoff shape. Draw all of its lines of symmetry.

147

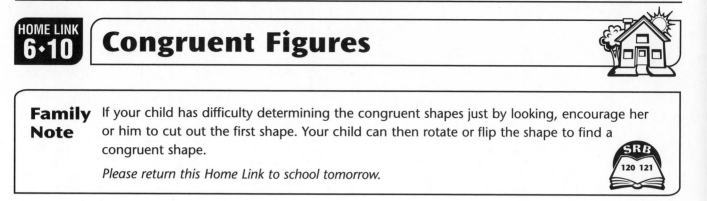

Congruent Figures

Two figures that are exactly the same size and shape are called **congruent** figures. In each of the following, circle the shape or shapes that are congruent to the first shape. Explain to someone at home why the other shape or shapes are *not* congruent to the first.

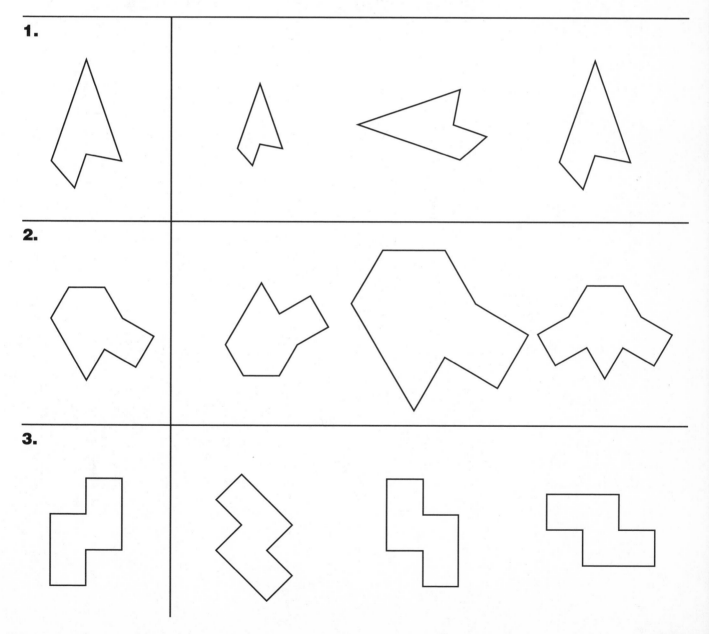

1.

2.

3.

**HOME LINK
6·11** **3-Dimensional Shapes**

> **Family Note** Have your child identify 3-dimensional shapes. Then help search for 3-D objects (or pictures of objects) around your home for your child to bring to school. Pages 112–119 in the *Student Reference Book* discuss 3-D shapes.
>
> *Please return this Home Link to school tomorrow.*
>
> **SRB**
> 112–119

1. Identify the pictures of the 3-dimensional shapes below.
   Use these words: *cone, prism, pyramid, cylinder,* and *sphere.*

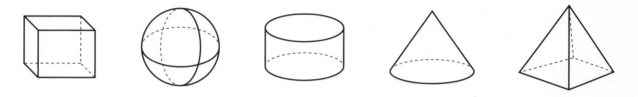

_____ _____ _____ _____ _____

2. Look around your home for objects or pictures of objects that are shaped like cones, prisms, pyramids, cylinders, and spheres. Ask someone at home if you may bring some of the objects or pictures to school to share with the class. Draw the shapes you find or write the names.

_____

_____

3. Explain to someone the differences between 2-dimensional (2-D) and 3-dimensional (3-D) shapes.

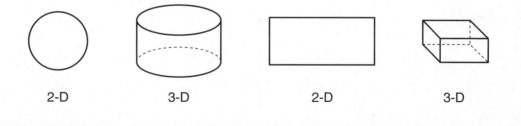

2-D   3-D   2-D   3-D

Practice

Write the problems on the back of this page. Solve.

4. $463 - 84 =$ _____   **5.** $54 - 29 =$ _____

151

HOME LINK 6·12 | Making a Solid Shape

> **Family Note** Our class has been exploring the characteristics and parts of various 3-dimensional shapes—especially prisms. The pattern on this page can be used to make one type of prism. Prisms are named for the shapes of their bases.
>
> *Please return this Home Link to school tomorrow.*

SRB
117

Cut on the dashed lines. Fold on the dotted lines. Tape or paste each TAB inside or outside the shape.

Discuss the following questions with someone at home:

1. What is this 3-D shape called? _____

2. What is the shape of the bases? _____

3. What is the shape of the other faces? _____

4. How many edges does the shape have? _____

5. How many vertices does the shape have? _____

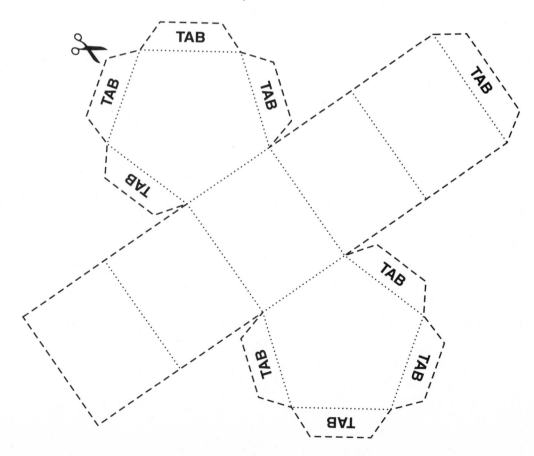

Unit 7: Family Letter

Multiplication and Division

In Unit 7, children will focus on learning the multiplication and division facts. Many of the same strategies that were used in previous grades for addition and subtraction will also be used for multiplication and division.

Children will review multiplication by 0, by 1, and by 10; multiplication facts having square products, such as $5 \times 5 = 25$ and $2 \times 2 = 4$; and the turn-around rule, which shows that $2 \times 5 = 10$ is the same as $5 \times 2 = 10$.

Children will also continue to work with fact families and Fact Triangles as they learn the multiplication and division facts.

$$7 \times 8 = 56$$
$$8 \times 7 = 56$$
$$56 \div 7 = 8$$
$$56 \div 8 = 7$$

Fact family for the
numbers 7, 8, and 56

Fact Triangle

The goal is for children to demonstrate automaticity with $\times 0$, $\times 1$, $\times 2$, $\times 5$, and $\times 10$ multiplication facts and to use strategies to compute remaining facts up to 10×10 by the end of the year.

Please keep this Family Letter for reference as your child works through Unit 7.

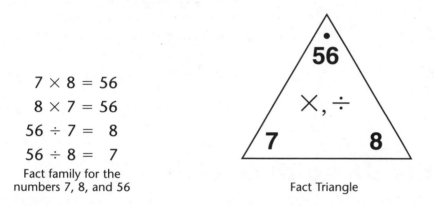

Vocabulary

Important terms in Unit 7:

factor Each of 2 or more numbers in a product. For example, $4 \times 3 = 12$; so 12 is the product, and 4 and 3 are the factors.

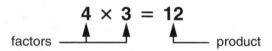

$$4 \times 3 = 12$$

factors ———product

product The result of multiplying 2 numbers, called factors. For example, in $4 \times 3 = 12$, the product is 12.

square number The product of a counting number and itself. For example, 25 is a square number, because $5 \times 5 = 25$.

estimate (1) An answer close to, or approximating, an exact answer. (2) To make an estimate.

parentheses ( ) Grouping symbols used to indicate which parts of an expression should be done first.

extended multiplication fact A multiplication fact involving multiples of 10, 100, and so on. In an extended multiplication fact, each factor has only one digit that is not 0. For example, 60×7, 70×6, and 60×70 are extended facts.

Building Skills through Games

In Unit 7, your child will practice multiplication and division skills by playing the following games. For detailed instructions, see the *Student Reference Book.*

Baseball Multiplication

Players use multiplication facts to score runs. Team members take turns pitching by rolling two dice to get two factors. Then players on the batting team take turns multiplying the two factors and saying the product.

Multiplication Bingo

Players take turns calling out the product of two numbers. If that number appears on their *Multiplication Bingo* cards, they put a penny on that number. The first player to get 4 pennies in a row, column, or diagonal calls out "Bingo!" and wins the game.

Name That Number

Players turn over a card to find a number they must rename using any combination of five faceup cards. They may add, subtract, multiply, or divide the numbers on 2 or more of the 5 cards that are number-side up.

The number 15 can be renamed using 3 cards as $3 \times 7 = 21$

$$21 - 6 = 15$$

Do-Anytime Activities

To work with your child on the concepts taught in this and previous units, try these interesting and rewarding activities:

1. Practice multiplication facts by playing games and by working with Fact Triangles.

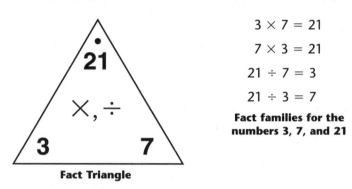

$3 \times 7 = 21$

$7 \times 3 = 21$

$21 \div 7 = 3$

$21 \div 3 = 7$

Fact families for the numbers 3, 7, and 21

Fact Triangle

2. Ask your child to count by certain intervals.
For example: Start at zero and count by 6s.

3. Provide your child with problems with missing factors for multiplication practice.
For example: 6 times what number equals 18?

4. Ask your child to estimate costs at the store.
For example: One loaf of bread costs $1.49. Two loaves are about $3.00.

5. Ask questions that involve equal sharing.
For example: Eight children share 64 paperback books. How many books does each child get?

6. Ask questions that involve equal groups.
For example: Pencils are packaged in boxes of 8. There are 3 boxes. How many pencils are there in all?

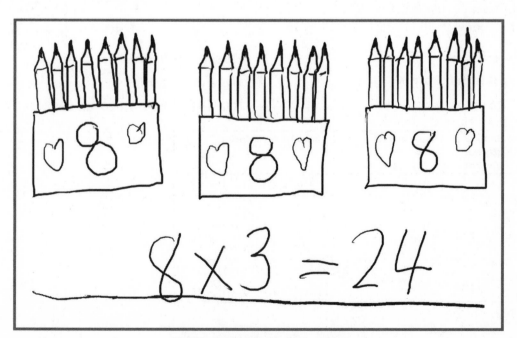

Child's drawing of equal groups

157

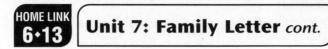

As You Help Your Child with Homework

As your child brings home assignments, you may want to go over the instructions together, clarifying them as necessary. The answers listed below will guide you through this unit's Home Links.

Home Link 7·2

1.

Factor	Factor	Product
3	5	15
7	2	14
4	10	40
8	8	64
9	5	45
864	1	864
10	10	100
0	999	0
1	48	48
243	0	0

5. 14,189

6. 3,166

Home Link 7·4

1a. $(17 - 10) + 3 = 10$ **1b.** $17 - (10 + 3) = 4$

2a. $(26 - 7) \times 2 = 38$ **2b.** $26 - (7 \times 2) = 12$

3a. $(24 - 17) - 6 = 1$ **3b.** $24 - (17 - 6) = 13$

4a. $3 \times (6 + 13) = 57$ **4b.** $(3 \times 6) + 13 = 31$

7. The parentheses are placed incorrectly.
The number model should be $(8 \times 4) + 4 = 36$.

Home Link 7·5

Scoring 15 Basketball Points

Home Link 7·6

1. $8 \times 200 = 1,600$ **2.** $9 \times 30 = 270$

$200 \times 8 = 1,600$ $30 \times 9 = 270$

$1,600 \div 8 = 200$ $270 \div 9 = 30$

$1,600 \div 200 = 8$ $270 \div 30 = 9$

3. $6 \times 40 = 240$

$40 \times 6 = 240$

$240 \div 6 = 40$

$240 \div 40 = 6$

Home Link 7·7

2. b. 1,750   **c.** 1,251   **f.** 545   **g.** 614

i. 522

Home Link 7·8

5. a. 1,200   **b.** 1,400   **c.** 400   **d.** 800

e. 2,000   **f.** 200   **g.** 2,000   **h.** 1,000

i. 0   Total = 9,000

Sample answers:

6. a. 10×10 **b.** 3×50

c. 30×3 **d.** 40×4

a 100	+ **b** 150	= 250
c 90	+ **d** 160	= 250

Total 500

Home Link 7·9

Mystery Numbers:

100; 199; 70; 44; 1,000; and 998

Number of 3-point baskets	Number of 2-point baskets	Number of 1-point baskets	Number models
5	0	0	$(5 \times 3) + (0 \times 2) + (0 \times 1) = 15$
0	5	5	$(0 \times 3) + (5 \times 2) + (5 \times 1) = 15$
3	3	0	$(3 \times 3) + (3 \times 2) + (0 \times 1) = 15$
4	0	3	$(4 \times 3) + (0 \times 2) + (3 \times 1) = 15$
2	3	3	$(2 \times 3) + (3 \times 2) + (3 \times 1) = 15$
1	6	0	$(1 \times 3) + (6 \times 2) + (0 \times 1) = 15$

1. 186 **2.** 509 **3.** 24

HOME LINK 7·1

Which Way Out?

Family Note   Today your child explored patterns in square products, such as 3 × 3 and 4 × 4. The activity below provides practice in identifying square products. Have your child start at the picture of the Minotaur and use a pencil so he or she can erase wrong turns. If it would be helpful, suggest that your child mark each square product before attempting to find a path.

Please return this Home Link to school tomorrow.

SRB 199

According to Greek mythology, there was a monster called the Minotaur that was half bull and half human. The king had a special mazelike dwelling built, from which the Minotaur could not escape. The dwelling, called a **labyrinth** (la buh rinth), had many rooms and passageways that formed a puzzle. Whoever went in could not find their way out without help. One day, a Greek hero, Theseus, decided to slay the monster. To find his way out of the labyrinth, his friend Ariadne gave him a very, very long ball of string to unwind as he walked through the passageways. After Theseus slew the Minotaur, he followed the string to escape.

Pretend you are Theseus. To find your way out, you may go through only those rooms numbered with square products. Start at the Minotaur's chambers and draw a path to the exit.

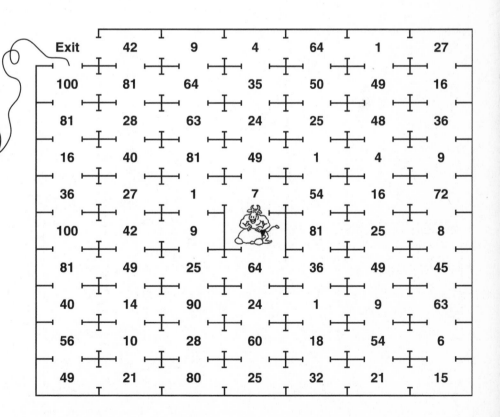

HOME LINK 7·2 **Factors and Products**

> **Family Note** Listen to your child explain what factors and products are before he or she writes the answers in the table. Then listen as your child tells you what he or she knows about multiplying by 1, multiplying by 0, and multiplying square numbers. Fact Triangles for the remaining multiplication/division facts are included with this Home Link.
>
> *Please return this Home Link to school tomorrow.*

1. Explain to someone at home what factors and products are. Find the missing products and factors in the table.

2. Write what you know about the products when you multiply by 1.

_____

_____

3. Write what you know about the products when you multiply by 0.

_____

4. Write what you know about facts with square numbers.

_____

Factor	Factor	Product
3	5	15
7		14
4	10	
8	8	
9		45
864	1	864
10		100
0	999	
	48	48
243		0

_____

(**Practice**)

Write these problems on the back of this page. Make a ballpark estimate for each. Solve. Show your work.

5.   7,201
   + 6,988

6.   3,623
   −   457

| **Unit** |
| |

ballpark estimate

_____

ballpark estimate

_____

161

HOME LINK 7·3

Multiplication Bingo (Easy Facts)

Family Note   Today the class learned to play *Multiplication Bingo*. This game is a good way to practice the multiplication facts. Ask your child to show you how to play the game; then play a couple of games. When your child is ready to practice harder facts, use the cards and list of numbers on the next page. Encourage your child to keep a record of the facts he or she misses.

Keep this Home Link at home.

Materials   ☐ number cards 1–6 and 10 (4 of each)

☐ 8 pennies or other counters for each player

☐ game mat for each player

Players   2 or 3

Directions

1. Write each of the following numbers in any order in one of the squares on a game mat: 1, 4, 6, 8, 9, 12, 15, 16, 18, 20, 24, 25, 30, 36, 50, 100.

2. Shuffle the number cards. Place the cards facedown on the table.

3. Take turns. When it is your turn, take the top 2 cards and call out the product of the 2 numbers. If the other players do not agree with your answer, check it using a calculator.

4. If your answer is correct and the product is a number on your grid, place a penny or a counter on that number.

5. If your answer is incorrect, you lose your turn.

6. The first player to get 4 counters in a row, column, or diagonal or 8 counters on the game mat calls out *Bingo!* and wins the game.

   If all the cards are used before someone wins, shuffle the cards again and keep playing.

HOME LINK 7·3 | Multiplication Bingo (All Facts)

Follow the same rules as for *Multiplication Bingo,* with the following exceptions:

◆ Use a deck of number cards with 4 cards each for the numbers 2 through 9.

◆ Write each of the numbers in the list in one of the squares on the grid. Don't write the numbers in order.

List of numbers

24	35	48	63
27	36	49	64
28	42	54	72
32	45	56	81

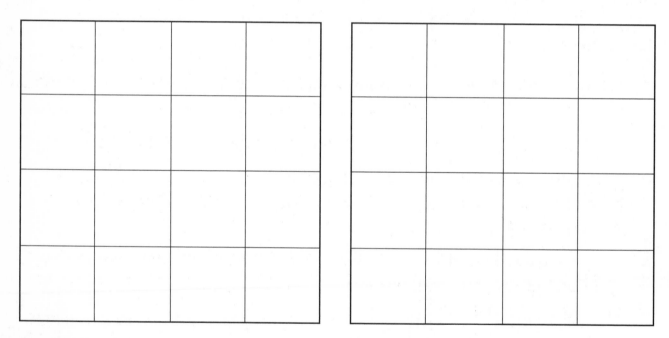

Record the facts you miss. Be sure to practice them.

_____ _____ _____

_____ _____ _____

_____ _____ _____

_____ _____ _____

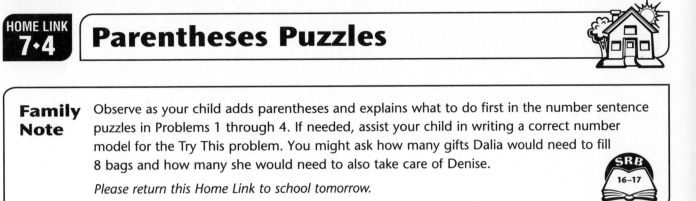

HOME LINK 7·4 | Parentheses Puzzles

Family Note Observe as your child adds parentheses and explains what to do first in the number sentence puzzles in Problems 1 through 4. If needed, assist your child in writing a correct number model for the Try This problem. You might ask how many gifts Dalia would need to fill 8 bags and how many she would need to also take care of Denise.

Please return this Home Link to school tomorrow.

SRB
16–17

Show someone at home how to add parentheses to complete the number sentences below. Remember that the parentheses are used to show what you do first.

1 a. $17 - 10 + 3 = 10$

1 b. $17 - 10 + 3 = 4$

2 a. $26 - 7 \times 2 = 38$

2 b. $26 - 7 \times 2 = 12$

3 a. $24 - 17 - 6 = 1$

3 b. $24 - 17 - 6 = 13$

4 a. $3 \times 6 + 13 = 57$

4 b. $3 \times 6 + 13 = 31$

Make up other parentheses puzzles below.

5 a. _____

5 b. _____

6 a. _____

6 b. _____

Try This

7. Dalia made 8 party bags for her birthday party. Each bag contained 4 small gifts for her friends. When Denise said that she could come, Dalia had to make one more bag with 4 gifts. How many small gifts did Dalia need to fill her bags?

Walter wrote this number model: $8 \times (4 + 4) = 64$
Explain Walter's mistake.

_____

_____

_____

HOME LINK 7·5 | Basketball Math

Family Note We have been using points scored in basketball to illustrate the use of parentheses in number models. Work with your child to find various combinations of 3-point, 2-point, and 1-point baskets that add up to 15 points. Ask your child to explain what the parentheses in the number models tell you about how to find the answers.

Please return this Home Link to school tomorrow.

SRB
16 17

Tell someone at home how basketball players can shoot baskets worth
3 points, 2 points, and 1 point. Find different ways a player can score 15 points.

Scoring 15 Basketball Points			
3 points	**2 points**	**1 point**	**Number Models**
3	*2*	*2*	$(3 \times 3) + (2 \times 2) + (2 \times 1) = 15$

Practice

Solve. Show your work.

Unit

1. 274
   − 88

2. 576
   − 67

3. 711
   − 687

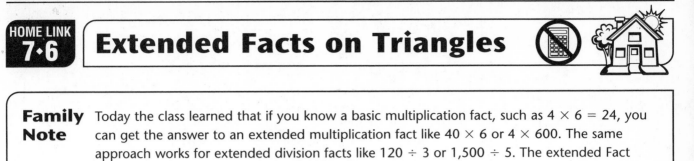

HOME LINK 7·6

Extended Facts on Triangles

Family Note Today the class learned that if you know a basic multiplication fact, such as 4 × 6 = 24, you can get the answer to an extended multiplication fact like 40 × 6 or 4 × 600. The same approach works for extended division facts like 120 ÷ 3 or 1,500 ÷ 5. The extended Fact Triangles on this page work the same way as the basic Fact Triangles.

Please return this Home Link to school tomorrow.

Fill in the extended Fact Triangles. Write the fact families.

1.

1,600
×, ÷
8 _____

_____ × _____ = _____

_____ × _____ = _____

_____ ÷ _____ = _____

_____ ÷ _____ = _____

2.

_____
×, ÷
9   30

_____ × _____ = _____

_____ × _____ = _____

_____ ÷ _____ = _____

_____ ÷ _____ = _____

3.

240
×, ÷
6   _____

_____ × _____ = _____

_____ × _____ = _____

_____ ÷ _____ = _____

_____ ÷ _____ = _____

4. Write your own.

_____
×, ÷
_____   _____

_____ × _____ = _____

_____ × _____ = _____

_____ ÷ _____ = _____

_____ ÷ _____ = _____

169

HOME LINK 7·7

Estimation

For each problem, first estimate whether the sum is greater than 500 or less than 500; then circle the correct comparison. Next give an exact result only to those problems with sums greater than 500.

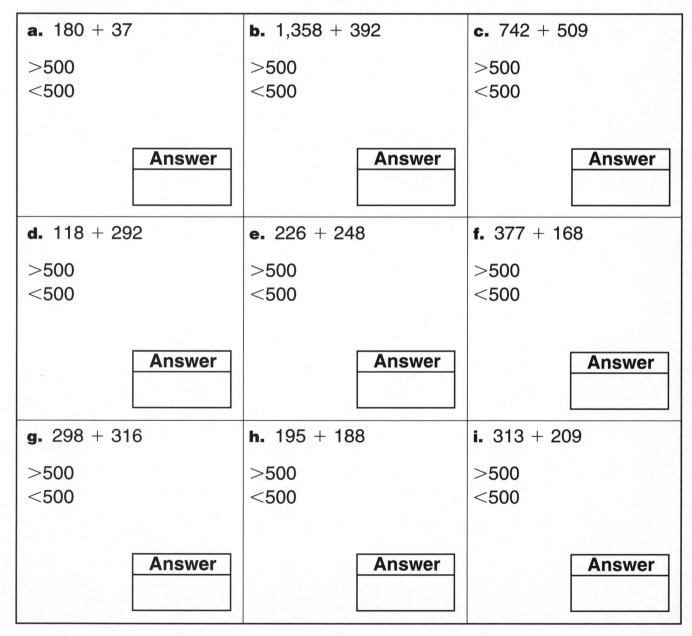

a. 180 + 37	b. 1,358 + 392	c. 742 + 509
>500 <500	>500 <500	>500 <500
Answer	**Answer**	**Answer**
d. 118 + 292	e. 226 + 248	f. 377 + 168
>500 <500	>500 <500	>500 <500
Answer	**Answer**	**Answer**
g. 298 + 316	h. 195 + 188	i. 313 + 209
>500 <500	>500 <500	>500 <500
Answer	**Answer**	**Answer**

HOME LINK 7·8

A Multiplication Puzzle

Family Note Practice finding products like 4 × 70, 900 × 5, and 30 × 50 with your child before he or she works the two puzzles.

Please return this Home Link to school tomorrow.

Work with someone at home.

1. Find each product below (for Problems 5a through 5i).

2. Record each product in the box labeled with the letter of the problem. For example, write the product for Problem **a** in Box **a**.

3. Add the numbers in each row. Write the sum next to the row.

4. Add these sums and write the answer in the Total box.

5. The number in the Total box should equal 3 × 3,000.

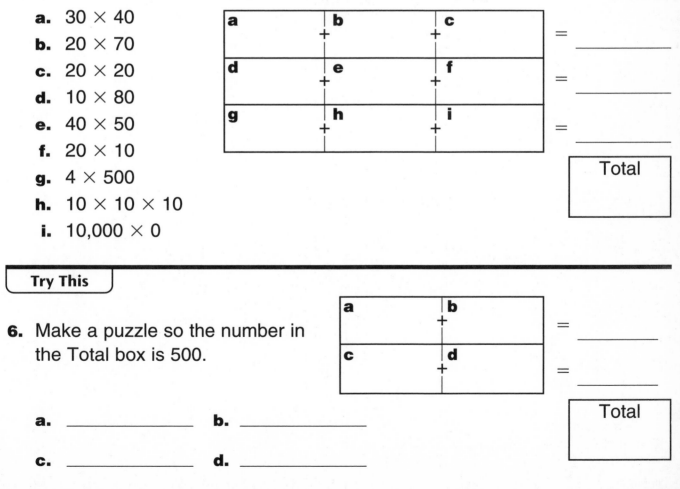

a. 30 × 40
b. 20 × 70
c. 20 × 20
d. 10 × 80
e. 40 × 50
f. 20 × 10
g. 4 × 500
h. 10 × 10 × 10
i. 10,000 × 0

Try This

6. Make a puzzle so the number in the Total box is 500.

a. _____ **b.** _____

c. _____ **d.** _____

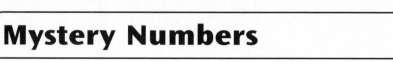

Mystery Numbers

> **Family Note** Help your child find each missing number by using all the clues. Then help your child create more clues for two other mystery numbers.
>
> *Please return this Home Link to school tomorrow.*

Find each missing number. Here are your clues.

Greater Than	Less Than	More Clues	Mystery Number
20	101	a 3-digit number	
197	200	any odd number	
67	80	has a zero in the ones place	
40	50	has the same digit in the tens place and the ones place	
917	1,072	has the same digit in the ones, tens, and hundreds places; has 4 digits	
996	1,015	a 3-digit even number	

Make up mystery-number puzzles. Write some clues and ask someone to find the numbers.

Greater Than	Less Than	More Clues	Mystery Number

HOME LINK 7·10

Unit 8: Family Letter

Fractions

Unit 8 has two primary objectives:

◆ to review the uses of fractions and
fraction notation

◆ to help children develop a solid understanding of
equivalent fractions, or fractions that have
the same value

The second objective is especially important, because
understanding equivalent fractions will help children
compare fractions and, later, calculate with fractions.

Children will build their understanding of equivalent
fractions by working with Fraction Cards and
name-collection boxes. Fraction Cards are shaded to
show a variety of fractions.

Name-collection boxes contain equivalent names for the
same number. For example, a $\frac{1}{2}$ name-collection
box can contain fractions such as $\frac{2}{4}$, $\frac{3}{6}$, and $\frac{4}{8}$ and the
decimal 0.50.

Children will also generate lists of equivalent fractions by
folding circles and rectangles into different numbers of equal parts.

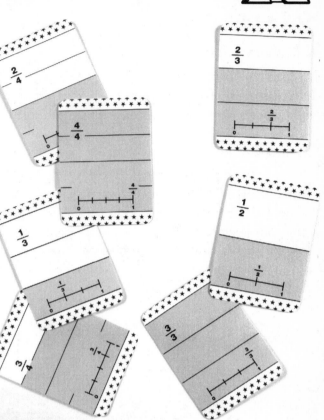

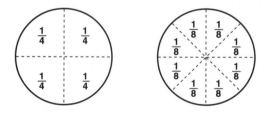

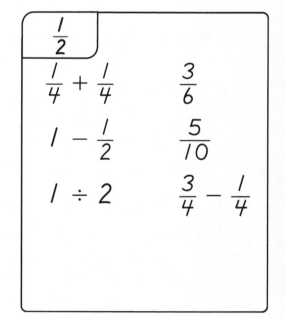

Throughout this unit, children will make up and solve
number stories involving fractions in everyday contexts.
They will solve number stories about collections of
real-world objects such as crayons, books, and cookies.

Finally, children will begin to name quantities greater than
1 with fractions such as $\frac{3}{2}$ and $\frac{5}{4}$ and with mixed numbers
such as $2\frac{1}{3}$.

**Please keep this Family Letter for reference as
your child works through Unit 8.**

177

HOME LINK 7·10 | **Unit 8: Family Letter** *cont.*

Vocabulary

Important terms in Unit 8:

fraction A number in the form $\frac{a}{b}$ where a and b are whole numbers and b is not 0. A fraction may be used to name part of a whole, to compare two quantities, or to represent division. For example, $\frac{2}{3}$ can be thought of as 2 divided by 3.

denominator The number below the line in a fraction. A fraction may be used to name part of a whole. If the whole is divided into equal parts, the denominator represents the number of equal parts into which the whole (the ONE or unit whole) is divided. In the fraction $\frac{a}{b}$, b is the denominator.

numerator The number above the line in a fraction. A fraction may be used to name part of a whole. If the whole (the ONE or unit whole) is divided into equal parts, the numerator represents the number of equal parts being considered. In the fraction $\frac{a}{b}$, a is the numerator.

equivalent fractions Fractions with different denominators that name the same number. For example, $\frac{1}{2}$ and $\frac{4}{8}$ are equivalent fractions.

mixed number A number that is written using both a whole number and a fraction. For example, $2\frac{1}{4}$ is a mixed number equal to $2 + \frac{1}{4}$.

numerator   **3** ← number of parts shaded
denominator   **4** ← number of equal parts

Building Skills through Games

In Unit 8, your child will practice multiplication skills, build his or her understanding of fractions, and practice skills related to chance and probability by playing the following games. For detailed instructions, see the *Student Reference Book*.

Baseball Multiplication

Players use multiplication facts to score runs. Team members take turns pitching by rolling two dice to get two factors. Then players on the batting team take turns multiplying the two factors and saying the product.

Equivalent Fractions Game

Players take turns turning over Fraction Cards and try to find matching cards that show equivalent fractions.

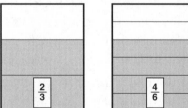

Fraction Top-It

Players turn over two Fraction Cards and compare the shaded parts of the cards. The player with the larger fraction keeps all the cards. The player with more cards at the end wins!

The Block-Drawing Game

Without letting the other players see the blocks, a Director puts five blocks in a paper bag and tells the players how many blocks are in the bag. A player takes a block out of the bag. The Director records the color of the block for all players to see. The player replaces the block. At any time, a player may say *Stop!* and guess how many blocks of each color are in the bag.

Unit 8: Family Letter *cont.*

Do-Anytime Activities

To work with your child on the concepts taught in this unit and in previous units, try these interesting and rewarding activities:

1. Help your child find fractions in the everyday world—in advertisements, on measuring tools, in recipes, and so on.

2. Count together by a 1-digit number. For example, start at 0 and count by 7s.

3. Dictate 5-, 6-, and 7-digit numbers for your child to write, such as *thirteen thousand, two hundred forty-seven* (13,247) and *three million, two hundred twenty-nine thousand, eight hundred fifty-six* (3,229,856). Also, write 5-, 6-, and 7- digit numbers for your child to read to you.

4. Practice extended multiplication and division facts such as $3 \times 7 =$ __, $30 \times 7 =$ __, and $300 \times 7 =$ __, and $18 \div 6 =$ __, $180 \div 6 =$ __, and $1{,}800 \div 6 =$ __.

As You Help Your Child with Homework

As your child brings home assignments, you may want to go over the instructions together, clarifying them as necessary. The answers listed below will guide you through this unit's Home Links.

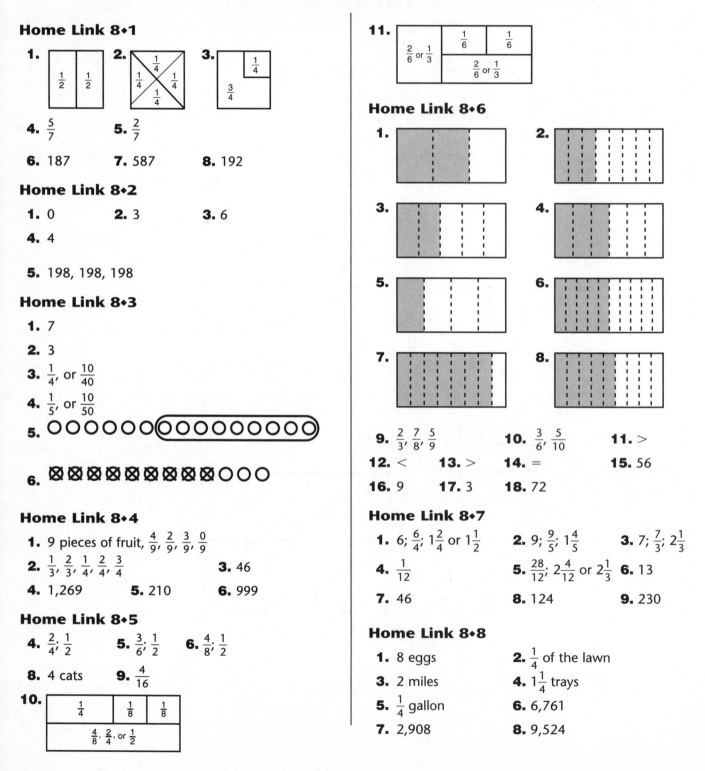

Home Link 8·1

1. [two rectangles each $\frac{1}{2}$]
2. [square divided into four triangles each $\frac{1}{4}$]
3. [square with $\frac{1}{4}$ and $\frac{3}{4}$]

4. $\frac{5}{7}$ 5. $\frac{2}{7}$

6. 187   7. 587   8. 192

Home Link 8·2

1. 0   2. 3   3. 6

4. 4

5. 198, 198, 198

Home Link 8·3

1. 7

2. 3

3. $\frac{1}{4}$, or $\frac{10}{40}$

4. $\frac{1}{5}$, or $\frac{10}{50}$

5. [circles with last 11 circled]

6. [X-filled circles, 9 filled then 3 empty]

Home Link 8·4

1. 9 pieces of fruit, $\frac{4}{9}, \frac{2}{9}, \frac{3}{9}, \frac{0}{9}$

2. $\frac{1}{3}, \frac{2}{3}, \frac{1}{4}, \frac{2}{4}, \frac{3}{4}$ 3. 46

4. 1,269   5. 210   6. 999

Home Link 8·5

4. $\frac{2}{4}, \frac{1}{2}$ 5. $\frac{3}{6}, \frac{1}{2}$ 6. $\frac{4}{8}, \frac{1}{2}$

8. 4 cats   9. $\frac{4}{16}$

10. [rectangle with $\frac{1}{4}, \frac{1}{8}, \frac{1}{8}$ and $\frac{4}{8}, \frac{2}{4},$ or $\frac{1}{2}$]

11. [rectangle with $\frac{2}{6}$ or $\frac{1}{3}$, $\frac{1}{6}$, $\frac{1}{6}$, $\frac{2}{6}$ or $\frac{1}{3}$]

Home Link 8·6

1. [shaded bar]   2. [shaded bar]

3. [shaded bar]   4. [shaded bar]

5. [shaded bar]   6. [shaded bar]

7. [shaded bar]   8. [shaded bar]

9. $\frac{2}{3}, \frac{7}{8}, \frac{5}{9}$ 10. $\frac{3}{6}, \frac{5}{10}$ 11. >

12. <   13. >   14. =   15. 56

16. 9   17. 3   18. 72

Home Link 8·7

1. 6; $\frac{6}{4}$; $1\frac{2}{4}$ or $1\frac{1}{2}$ 2. 9; $\frac{9}{5}$; $1\frac{4}{5}$ 3. 7; $\frac{7}{3}$; $2\frac{1}{3}$

4. $\frac{1}{12}$ 5. $\frac{28}{12}$; $2\frac{4}{12}$ or $2\frac{1}{3}$ 6. 13

7. 46   8. 124   9. 230

Home Link 8·8

1. 8 eggs   2. $\frac{1}{4}$ of the lawn

3. 2 miles   4. $1\frac{1}{4}$ trays

5. $\frac{1}{4}$ gallon   6. 6,761

7. 2,908   8. 9,524

HOME LINK 8·1 — Fractions All Around

Each square flag below represents the ONE. Write the fractions that name each region inside each flag.

1. 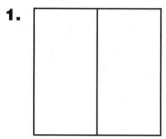   **2.** 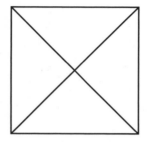   **3.**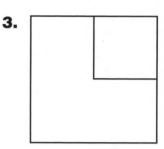

Write the fractions.

4. _____ of the buttons have 4 holes.

5. _____ of the buttons have 2 holes.

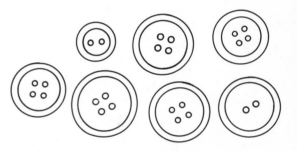

Look for items around your home that have fractions or decimals on them, such as recipes, measuring cups, wrenches, package labels, or pictures in newspapers. Ask permission to bring them to school to display in our Fractions Museum.

Practice

Unit

Solve. Show your work.

6.   275
    − 88

7.   684
    − 97

8.   429
   − 237

HOME LINK 8·2 Drawing Blocks

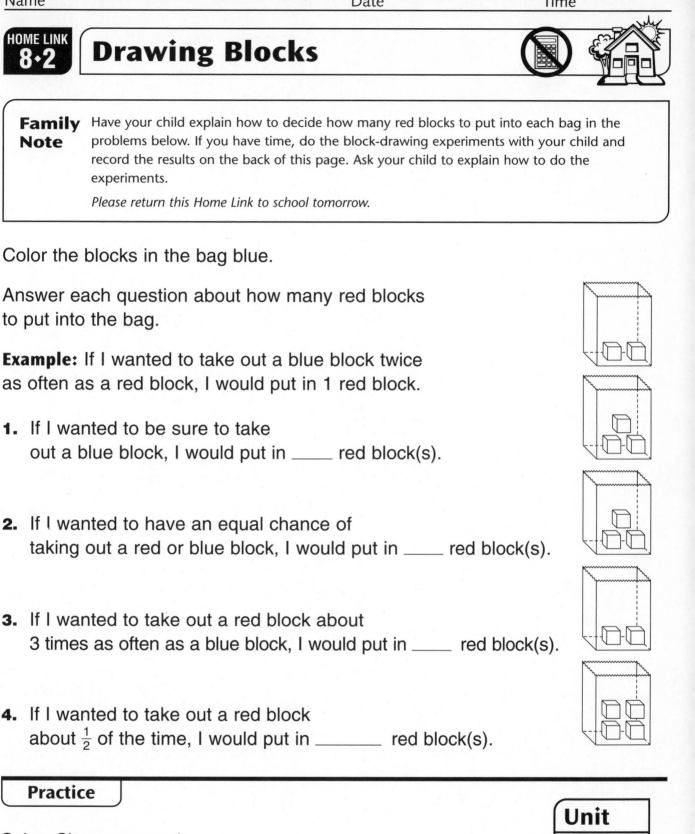

Family Note Have your child explain how to decide how many red blocks to put into each bag in the problems below. If you have time, do the block-drawing experiments with your child and record the results on the back of this page. Ask your child to explain how to do the experiments.

Please return this Home Link to school tomorrow.

Color the blocks in the bag blue.

Answer each question about how many red blocks to put into the bag.

Example: If I wanted to take out a blue block twice as often as a red block, I would put in 1 red block.

1. If I wanted to be sure to take out a blue block, I would put in _____ red block(s).

2. If I wanted to have an equal chance of taking out a red or blue block, I would put in _____ red block(s).

3. If I wanted to take out a red block about 3 times as often as a blue block, I would put in _____ red block(s).

4. If I wanted to take out a red block about $\frac{1}{2}$ of the time, I would put in _____ red block(s).

Practice

Solve. Show your work.

5. 765
   − 567

6. 987
   − 789

7. 432
   − 234

Unit

HOME LINK
8·3

Fraction Number Stories

Family Note Your child may benefit from modeling the number stories with pennies or counters. Help your child think about the problems as stories about equal shares or equal groups.

Please return this Home Link to school tomorrow.

Solve each problem. Tell someone at home how you did it.
Draw a picture on the back if it will help.

1. Lucy was playing a card game with 2 friends.
   They were playing with a deck of 21 cards.
   Lucy dealt $\frac{1}{3}$ of the deck to each person.
   How many cards did Lucy get? _____ cards

2. Jonathan bought 12 pencils. He gave $\frac{1}{2}$ of them to his brother
   and $\frac{1}{4}$ of them to his friend Mike.
   How many pencils did he give to Mike? _____ pencils

3. Gerard was reading a book with 40 pages.
   He read 10 pages in an hour.
   What fraction of the book did he read in an hour? _____

4. Melissa was reading a book with 50 pages.
   She read 10 pages in an hour.
   What fraction of the book did she read in an hour? _____

Follow the instructions below.

5. Draw 15 small circles. Circle $\frac{3}{5}$ of them.

6. Draw 12 small circles. Put an X through $\frac{3}{4}$ of them.

185

HOME LINK 8·4 | **Fraction Puzzles**

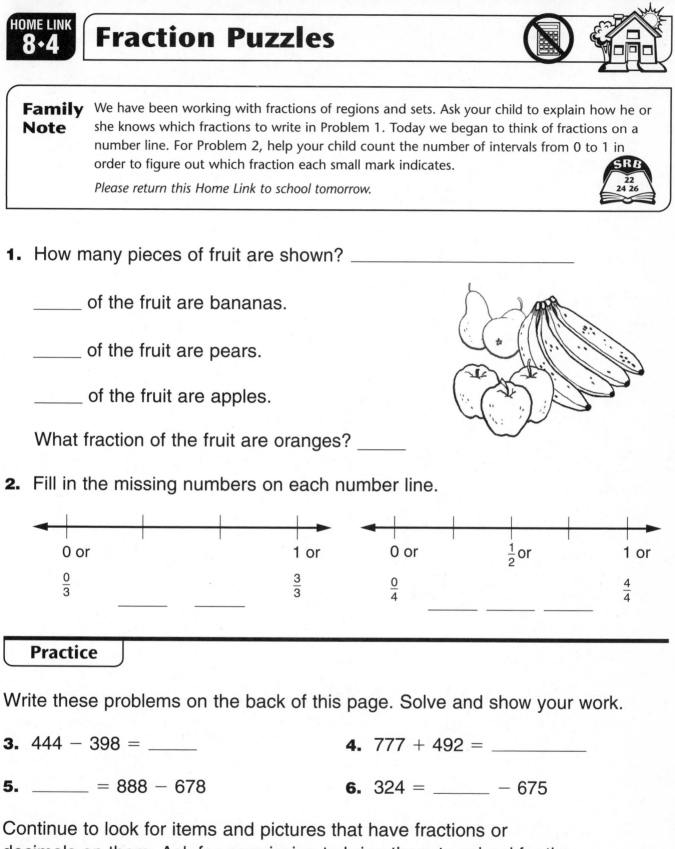

Family Note We have been working with fractions of regions and sets. Ask your child to explain how he or she knows which fractions to write in Problem 1. Today we began to think of fractions on a number line. For Problem 2, help your child count the number of intervals from 0 to 1 in order to figure out which fraction each small mark indicates.

Please return this Home Link to school tomorrow.

SRB
22
24 26

1. How many pieces of fruit are shown? _____

   _____ of the fruit are bananas.

   _____ of the fruit are pears.

   _____ of the fruit are apples.

   What fraction of the fruit are oranges? _____

2. Fill in the missing numbers on each number line.

   0 or $\frac{0}{3}$ _____ _____ 1 or $\frac{3}{3}$

   0 or $\frac{0}{4}$ $\frac{1}{2}$ or _____ _____ 1 or $\frac{4}{4}$

Practice

Write these problems on the back of this page. Solve and show your work.

3. 444 − 398 = _____

4. 777 + 492 = _____

5. _____ = 888 − 678

6. 324 = _____ − 675

Continue to look for items and pictures that have fractions or decimals on them. Ask for permission to bring them to school for the Fractions Museum.

187

HOME LINK 8·5

Equipment Equivalent Fractions

Family Note The class continues fraction work by finding equivalent names for fractions. Different fractions that name the same amount are called equivalent fractions. The fractions that complete Problems 4–6 are equivalent. If needed, help your child name the fractional parts in these problems. Ask your child to explain the fraction name she or he chooses in Problem 9—a fraction that is equivalent to $\frac{1}{4}$ and describes the fraction of cats circled.

SRB
27–30

Please return this Home Link to school tomorrow.

The pictures show three kinds of pie. Use a straightedge to do the following:

1. Divide the peach pie into 4 equal pieces. Shade 2 of the pieces.

2. Divide the blueberry pie into 6 equal pieces. Shade 3 of the pieces.

3. Divide the cherry pie into 8 equal pieces. Shade 4 of the pieces.

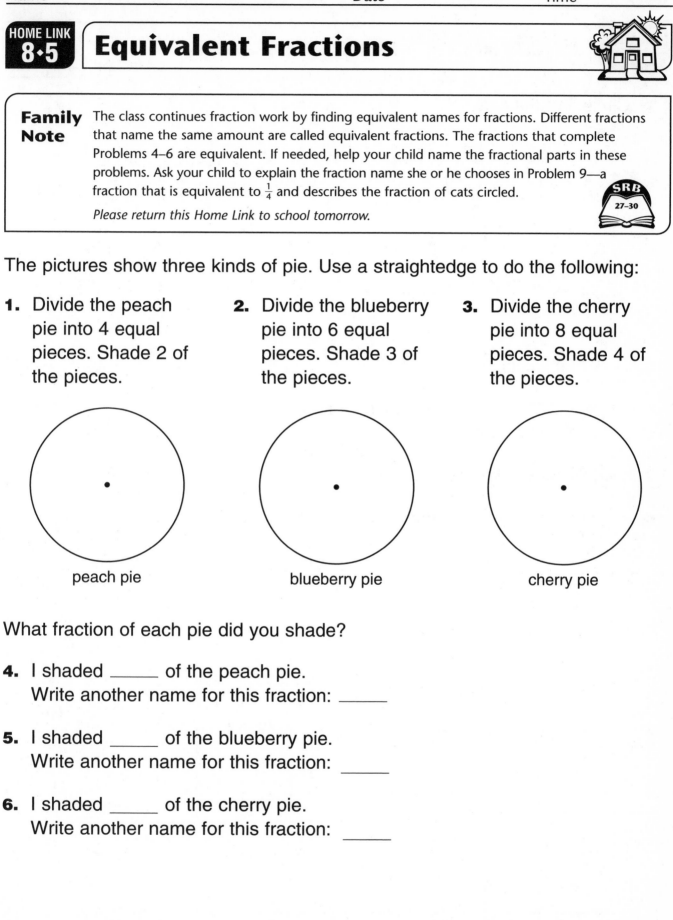

peach pie   blueberry pie   cherry pie

What fraction of each pie did you shade?

4. I shaded _____ of the peach pie.
Write another name for this fraction: _____

5. I shaded _____ of the blueberry pie.
Write another name for this fraction: _____

6. I shaded _____ of the cherry pie.
Write another name for this fraction: _____

HOME LINK 8·5 | **Equivalent Fractions** *continued*

7. Circle $\frac{1}{4}$ of the cats.

8. How many cats did you circle? _____

9. Write a fraction that describes the group of
cats you circled and that is equivalent to $\frac{1}{4}$. _____

Each whole rectangle below is ONE. Write a fraction inside each part.

10.

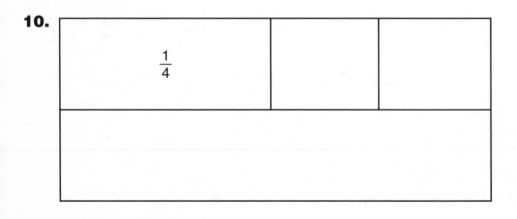

$\frac{1}{4}$

11.

$\frac{1}{6}$

HOME LINK 8·6 Comparing Fractions to $\frac{1}{2}$

Family Note Your child's class is comparing fractions to determine whether they are larger, smaller, or equal to $\frac{1}{2}$. Ask your child to explain how to tell which category a fraction fits into. For more on this topic, see *Student Reference Book* pages 13, 31, and 32.

Please return this Home Link to school tomorrow.

Shade each rectangle to match the fraction below it. **Example:** $\frac{2}{4}$

1. $\frac{2}{3}$

2. $\frac{3}{8}$

3. $\frac{2}{5}$

4. $\frac{3}{6}$

5. $\frac{1}{4}$

6. $\frac{5}{10}$

7. $\frac{7}{8}$

8. $\frac{5}{9}$

9. List the fractions above that are greater than $\frac{1}{2}$. _____

10. List the fractions above that are equal to $\frac{1}{2}$. _____

Insert $<$, $>$, or $=$ in each problem below. Draw pictures to help you.

11. $\frac{6}{8}$ _____ $\frac{1}{2}$

12. $\frac{2}{9}$ _____ $\frac{1}{2}$

13. $\frac{10}{12}$ _____ $\frac{1}{2}$

14. $\frac{6}{12}$ _____ $\frac{1}{2}$

$<$ means *is less than*
$>$ means *is greater than*
$=$ means *is equal to*

Practice

Solve.

15. $7 \times 8 =$ _____

16. $54 = 6 \times$ _____

17. $8 \times$ _____ $= 24$

18. $9 \times 8 =$ _____

HOME LINK 8·7 Fractions and Mixed Numbers

Family Note Today the class began looking at fractions greater than 1 and mixed numbers. We have been working with region or area models (shaded areas) for these numbers. Problem 5 asks about fractions of a set. The *whole* is a dozen eggs, so each egg is $\frac{1}{12}$ of the whole. Have your child explain how he or she figured out what the fraction and mixed number should be for the egg-carton drawings.

Please return this Home Link to school tomorrow.

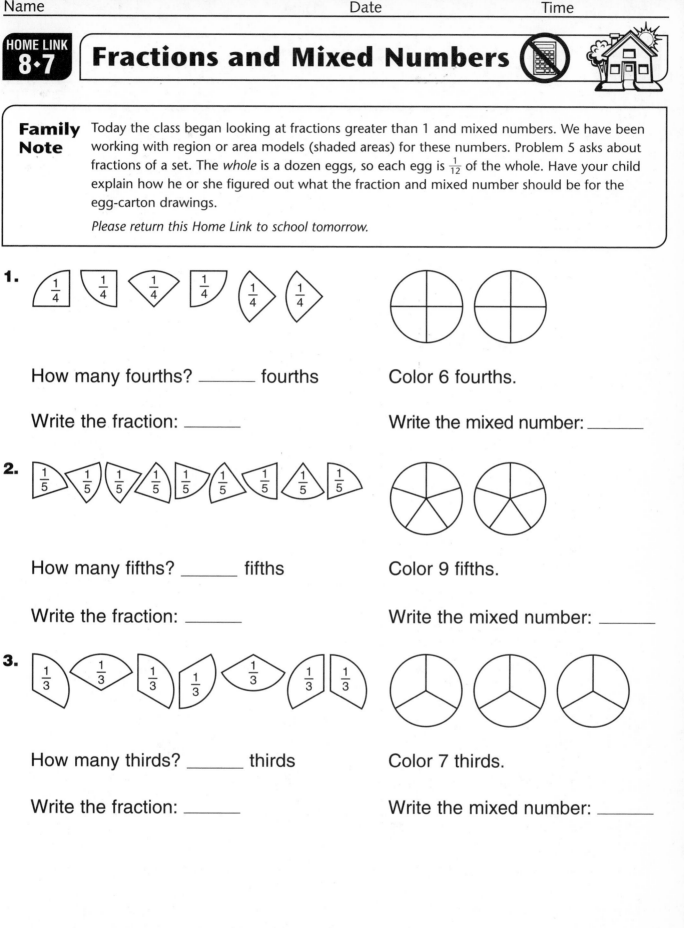

1.

How many fourths? _____ fourths

Write the fraction: _____

Color 6 fourths.

Write the mixed number: _____

2.

How many fifths? _____ fifths

Write the fraction: _____

Color 9 fifths.

Write the mixed number: _____

3.

How many thirds? _____ thirds

Write the fraction: _____

Color 7 thirds.

Write the mixed number: _____

HOME LINK 8·7 | **Fractions and Mixed Numbers** *cont.*

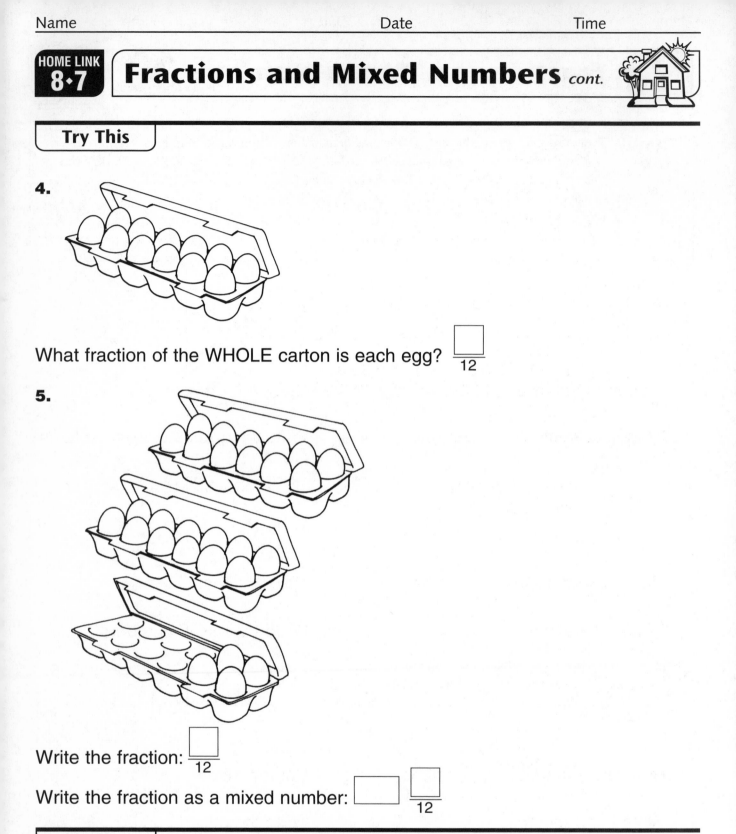

Try This

4.

What fraction of the WHOLE carton is each egg? $\dfrac{\boxed{}}{12}$

5.

Write the fraction: $\dfrac{\boxed{}}{12}$

Write the fraction as a mixed number: $\boxed{}\,\dfrac{\boxed{}}{12}$

Practice

Write these problems on the back of this page. Solve and show your work.

6.	301	**7.**	27	**8.**	600	**9.**	131
	− 288		+ 19		− 476		+ 99

HOME LINK 8·8

Fraction Number Stories

Family Note In class we have been solving many kinds of fraction number stories. If some of these Home Link problems seem difficult, encourage your child to model them with pennies or draw pictures to help solve them.

Please return this Home Link to school tomorrow.

SRB 22–24

Solve these fraction stories. Use pennies, counters, or pictures to help.

1. Elizabeth bought a dozen eggs. She dropped her bag on the way home, and $\frac{2}{3}$ of the eggs broke. How many eggs broke? _____ eggs

2. Katie mowed $\frac{3}{4}$ of the lawn before lunch. What fraction of the lawn did she have to finish after lunch? _____ of the lawn

3. Donnie lives 1 mile from school. One day he walked $\frac{1}{2}$ of the way to school when he remembered he had to return home to get a book. When he finally made it to school, how far did he walk in all? _____ miles

4. Sheridan made 4 trays of cookies. She took 2 trays to school for her classmates. She took $\frac{3}{4}$ of a tray of cookies to her teacher. How many trays of cookies did Sheridan have left? _____ trays

5. Jackson needed 2 pints of milk for his recipe. If he had one gallon of milk in the refrigerator, how much did he use? (*Hint:* 1 gallon = 4 quarts, and 1 quart = 2 pints) _____ gallon

---

Practice

Write these problems on the back of this page. Solve and show your work.

Unit

6. $2{,}083 + 4{,}678 =$ _____

7. $6{,}714 - 3{,}806 =$ _____

8. $4{,}762 + 4{,}762 =$ _____

Unit 9: Family Letter

HOME LINK 8·9

Multiplication and Division

In Unit 9, children will develop a variety of strategies for multiplying whole numbers. They will begin by using mental math (computation done by counting fingers, drawing pictures, making diagrams, and computing in one's head). Later in this unit, children will be introduced to two specific algorithms, or methods, for multiplication: the partial-products algorithm and the lattice method.

Partial-Products Algorithm

The partial-products algorithm is a variation of the traditional multiplication algorithm that most adults learned as children. Note that the multiplication is done from left to right and emphasizes place value in the numbers being multiplied.

$$
\begin{array}{r}
28 \\
\times\ \ 4 \\
\end{array}
$$

Multiply 4 × 20.	→ 80	First, calculate 4 [20s].
Multiply 4 × 8.	→ + 32	Then calculate 4 [8s].
Add the two partial products.	→ 112	Finally, add the two partial products.

It is important that when children verbalize this method, they understand and say *4 [20s]*, not *4 × 2*. In doing so, they gain a better understanding of the magnitude of numbers along with better number sense.

$$
\begin{array}{r}
379 \\
\times\ \ 4 \\
\end{array}
$$

Multiply 4 × 300.	→ 1,200	First, calculate 4 [300s].
Multiply 4 × 70.	→ 280	Second, calculate 4 [70s].
Multiply 4 × 9.	→ + 36	Then calculate 4 [9s].
Add the three partial products.	→ 1,516	Finally, add the three partial products.

Check that when your child is verbalizing this strategy, he or she says *4 [300s]*, not *4 × 3*; and *4 [70s]*, not *4 × 7*. Using this strategy will also help to reinforce your child's facility with the basic multiplication facts and their extensions.

Lattice Method

Third Grade Everyday Mathematics introduces the lattice method of multiplication for several reasons: This algorithm is historically interesting; it provides practice with multiplication facts and addition of 1-digit numbers; and it is fun. Also, some children find it easier to use than other methods of multiplication.

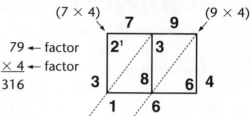

Step 1 Write the factors on the outside of the lattice. Line up one factor with the column(s); the other with the row(s).

Step 2 Multiply each digit in one factor by each digit in the other factor.

Step 3 Write each product in one small box; ones place digits in the bottom-right half; tens place digits in the upper-left half. When the product is a single-digit answer, write a zero in the upper-left half.

Step 4 Beginning on the right, add the numbers inside the lattice along each diagonal. If the sum on a diagonal exceeds 9, add the excess 10s in the next diagonal to the left.

The lattice method and the partial-products algorithm help prepare children for a division algorithm they will learn in fourth grade. Children will choose the algorithms that work best for them.

Also in this unit, children will…

◆ Write and solve multiplication and division number stories involving multiples of 10, 100, and 1,000.

◆ Solve division number stories and interpret the remainders.

◆ Increase their understanding of positive and negative numbers.

Vocabulary

Important terms in Unit 9:

algorithm A step-by-step set of instructions for doing something such as carrying out computation or solving a problem.

degree Celsius (°C) A unit for measuring temperature on the Celsius scale. 0°Celsius is the freezing point of water. 100°Celsius is the boiling point of water.

degree Fahrenheit (°F) A unit for measuring temperature on the Fahrenheit scale. 32°F is the freezing point of water. 212°F is the boiling point of water.

negative number A number less than or below zero; a number to the left of zero on a horizontal number line. The symbol − may be used to write a negative number. For example, negative 5 is usually written as −5.

positive number A number that is greater than zero; a number to the right of zero on a horizontal number line. A positive number may be written using the + symbol but it is usually written without it. For example, +10 = 10.

factor of a counting number *n* A counting number whose product with some other counting number equals *n*. For example, 2 and 3 are factors of 6 because 2 × 3 = 6. But 4 is not a factor of 6 because 4 × 1.5 = 6 and 1.5 is not a counting number.

Do-Anytime Activities

To work with your child on the concepts taught in this unit and in previous units, try these interesting and rewarding activities:

1. As the class proceeds through the unit, give your child multiplication problems related to the lessons covered, such as 9×23, 3×345, 20×65, and 43×56.

2. Continue to work on multiplication and division facts by using Fact Triangles and fact families, or by playing games.

3. Play *Baseball Multiplication, Factor Bingo,* and other games described in the *Student Reference Book.*

4. Write decimals for your child to read, such as 0.82 (eighty-two hundredths); 0.7 (seven tenths); 0.348 (three hundred forty-eight thousandths); and so on. Ask your child to identify digits in various places—the tenths place, hundredths place, thousandths place. Look for decimals in newspapers and on food containers.

5. Practice extended multiplication and division facts such as $3 \times 7 = ?$, $3 \times 70 = \_\_$, $3 \times 700 = \_\_$; $18 \div 6 = \_\_$, $180 \div 6 = \_\_$, and $1,800 \div 6 = \_\_$.

As You Help Your Child with Homework

As your child brings home assignments, you may want to go over the instructions together, clarifying them as necessary. The answers listed below will guide you through this unit's Home Links.

Home Link 9·1

1. 31   **2.** 25   **3.** 22

4. 13 or 18   **5.** 12 or 24   **6.** 56; 560; 5,600

7. 20; 200; 20,000

Home Link 9·2

1. a. 56; 56   **b.** 560; 560   **c.** 7

   d. 70   **e.** 8   **f.** 8

2. a. 63; 63   **b.** 630; 630   **c.** 7

   d. 70   **e.** 9   **f.** 9

3. a. 40; 40   **b.** 400; 400   **c.** 50

   d. 50   **e.** 8   **f.** 80

Home Link 9·3

1. 7 raccoons   **2.** 500 lb   **3.** 100 arctic foxes

4. 600 lb   **5.** 400 lb   **6.** 60 beluga whales

Home Link 9·4

1. 93   **2.** 375   **3.** 765

4. 258   **5.** 1,134

Home Link 9·5

1. yes; estimate; $0.80 \times 7 = \$5.60$

2. \$12.72; calculate; $\$2.12 \times 6 = \$12.72$

3. \$0.90; Sample answer: calculate; 10 cards is \$6.00 times 2. Compare that with \$1.29 times 10. Then subtract to find the difference.

Home Link 9·6

1 row: yes; 18 chairs   **7 rows:** no

2 rows: yes; 9 chairs   **8 rows:** no

3 rows: yes; 6 chairs   **9 rows:** yes; 2 chairs

4 rows: no   **10 rows:** no

5 rows: no   **18 rows:** yes; 1 chair

6 rows: yes; 3 chairs   1; 18; 2; 9; 3; 6

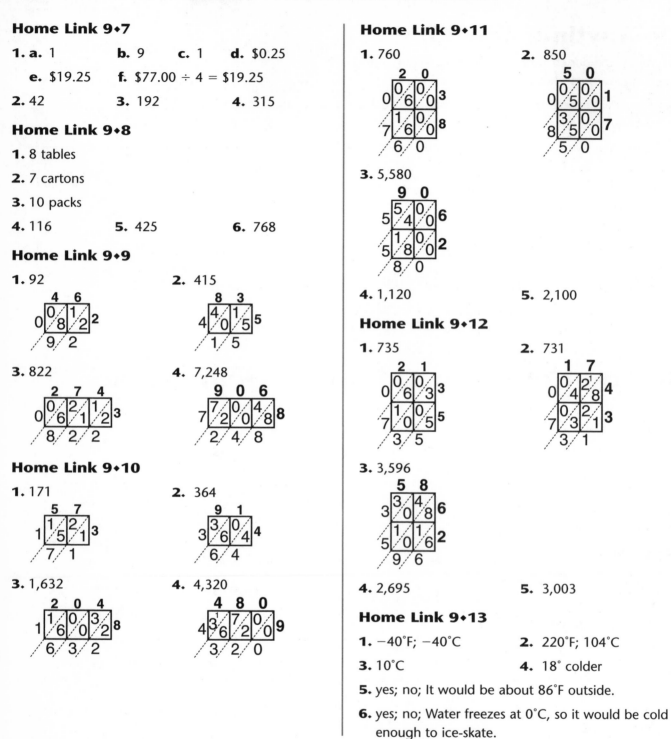

Home Link 9·7

1. a. 1   **b.** 9   **c.** 1   **d.** $0.25

   e. $19.25   **f.** $77.00 ÷ 4 = $19.25

2. 42   **3.** 192   **4.** 315

Home Link 9·8

1. 8 tables

2. 7 cartons

3. 10 packs

4. 116   **5.** 425   **6.** 768

Home Link 9·9

1. 92

2. 415

3. 822

4. 7,248

Home Link 9·10

1. 171

2. 364

3. 1,632

4. 4,320

Home Link 9·11

1. 760   **2.** 850

3. 5,580

4. 1,120   **5.** 2,100

Home Link 9·12

1. 735   **2.** 731

3. 3,596

4. 2,695   **5.** 3,003

Home Link 9·13

1. −40°F; −40°C   **2.** 220°F; 104°C

3. 10°C   **4.** 18° colder

5. yes; no; It would be about 86°F outside.

6. yes; no; Water freezes at 0°C, so it would be cold enough to ice-skate.

HOME LINK
9·1

Who Am I?

In each riddle, I am a different whole number. Use the clues to find out who I am.

1. **Clue 1:** I am greater than 30 and less than 40.   **Who am I?**
   Clue 2: The sum of my digits is less than 5.   _____

2. **Clue 1:** I am greater than 15 and less than 40.   **Who am I?**
   Clue 2: If you double me, I become
   a number that ends in 0.   _____
   Clue 3: $\frac{1}{5}$ of me is equal to 5.

3. **Clue 1:** I am less than 100.   **Who am I?**
   Clue 2: The sum of my digits is 4.   _____
   Clue 3: Half of me is an odd number.

4. **Clue 1:** If you multiply me by 2, I become   **Who am I?**
   a number greater than 20 and less than 40.   _____
   Clue 2: If you multiply me by 6, I end in 8.
   Clue 3: If you multiply me by 4, I end in 2.

5. **Clue 1:** Double my tens digit to get   **Who am I?**
   my ones digit.   _____
   Clue 2: Double me and I am less than 50.

Practice

Solve.

6. $8 \times 7 =$ _____   7. $5 \times 4 \; =$ _____

     $80 \times 7 =$ _____ $5 \times 40 \; =$ _____

    $800 \times 7 =$ _____ $50 \times 400 =$ _____

HOME LINK 9·2 Multiplication Facts and Extensions

Family Note Help your child practice multiplication facts and their extensions. Observe as your child creates fact extensions, demonstrating further understanding of multiplication.

Please return this Home Link to school tomorrow.

Solve each problem.

1. a. 8 [7s] = _____, or 8 × 7 = _____

   b. 8 [70s] = _____, or 8 × 70 = _____

   c. How many 8s in 56? _____   **d.** How many 8s in 560? _____

   e. How many 7s in 56? _____   **f.** How many 70s in 560? _____

2. a. 9 [7s] = _____, or 9 × 7 = _____

   b. 9 [70s] = _____, or 9 × 70 = _____

   c. How many 9s in 63? _____   **d.** How many 9s in 630? _____

   e. How many 7s in 63? _____   **f.** How many 70s in 630? _____

3. a. 8 [5s] = _____, or 8 × 5 = _____

   b. 8 [50s] = _____, or 8 × 50 = _____

   c. How many 8s in 400? _____   **d.** How many 80s in 4,000? _____

   e. How many 50s in 400? _____   **f.** How many 50s in 4,000? _____

4. Write a multiplication fact you are trying to learn.
Then use your fact to write some fact extensions like those above.

_____

_____

_____

_____

HOME LINK 9·3 | Multiplication Number Stories

Family Note Your child's class is beginning to solve multidigit multiplication and division problems. Although we have practiced multiplication and division with multiples of 10, we have been doing most of our calculating mentally. Encourage your child to explain a solution strategy for each of the problems below.

Please return this Home Link to school tomorrow.

SRB
250–253

1. How many 30-pound raccoons would weigh about as much as a 210-pound harp seal? _____

2. How much would an alligator weigh if it weighed 10 times as much as a 50-pound sea otter? _____

3. How many 20-pound arctic foxes would weigh about as much as a 2,000-pound beluga whale? _____

4. Each porcupine weighs 30 pounds. A black bear weighs as much as 20 porcupines. How much does the black bear weigh? _____

5. A bottle-nosed dolphin could weigh twice as much as a 200-pound common dolphin. How much could the bottle-nosed dolphin weigh? _____

Try This

6. How many 2,000-pound beluga whales would weigh as much as one 120,000-pound right whale? _____

205

HOME LINK 9·4 | The Partial-Products Algorithm

Use the partial-products algorithm to solve these problems:

Example	**1.**
$\begin{array}{r} 46 \\ \times\ 7 \\ \hline \end{array}$ 7 [40s]→ 280 7 [6s]→ + 42 280 + 42→ 322	$\begin{array}{r} 31 \\ \times\ 3 \\ \hline \end{array}$
2. $\begin{array}{r} 75 \\ \times\ 5 \\ \hline \end{array}$	**3.** $\begin{array}{r} 85 \\ \times\ 9 \\ \hline \end{array}$
4. $\begin{array}{r} 43 \\ \times\ 6 \\ \hline \end{array}$	**5.** $\begin{array}{r} 162 \\ \times\ \ \ 7 \\ \hline \end{array}$

HOME LINK 9·5 | **Saving at the Stock-Up Sale**

Family Note Today the class used mental math and the partial-products algorithm to solve shopping problems. Note that for some of the problems below, an estimate will answer the question. For others, an exact answer is needed. If your child is able to make the calculations mentally, encourage him or her to explain the solution strategy to you.

Please return this Home Link to school tomorrow.

SRB 250–253 191

Decide whether you will need to estimate or calculate an exact answer to solve each problem below. Then solve the problem and show what you did. Record the answer and write the number model (or models) you used.

1. Phil has $6.00. He wants to buy Creepy Creature erasers. They cost $1.05 each. If he buys more than 5, they are $0.79 each. Does he have enough money to buy 7 Creepy Creature erasers? _____

Number model: _____

2. Mrs. Katz is buying cookies for a school party. The cookies cost $2.48 per dozen. If she buys more than 4 dozen, they cost $2.12 per dozen. How much are 6 dozen? _____

Number model: _____

3. Baseball cards are on sale for $1.29 per card, or 5 cards for $6. Marty bought 10 cards. How much did he save with the special price? _____

Explain how you found your answer.

_____

_____

_____

209

HOME LINK 9·6 **Arrays and Factors**

Family Note Discuss with your child all the ways to arrange 18 chairs in equal rows. Then help your child use this information to list the factors of 18 (pairs of numbers whose product is 18).

Please return this Home Link to school tomorrow.

SRB 64–67

Work with someone at home.

The third-grade class is putting on a play. Children have invited 18 people. Gilda and Harvey are in charge of arranging the 18 chairs. They want to arrange them in rows with the same number of chairs in each row, with no chairs left over.

Yes or no: Can they arrange the chairs in …	If yes, how many chairs in each row?
1 row? _____	_____ chairs
2 rows? _____	_____ chairs
3 rows? _____	_____ chairs
4 rows? _____	_____ chairs
5 rows? _____	_____ chairs
6 rows? _____	_____ chairs
7 rows? _____	_____ chairs
8 rows? _____	_____ chairs
9 rows? _____	_____ chairs
10 rows? _____	_____ chairs
18 rows? _____	_____ chairs

List all the factors of the number 18. (*Hint:* 18 has exactly 6 factors.)

_____ _____ _____

_____ _____ _____

How does knowing all the ways to arrange 18 chairs in equal rows help you find the factors of 18? Tell someone at home.

_____

_____

_____

_____

_____

211

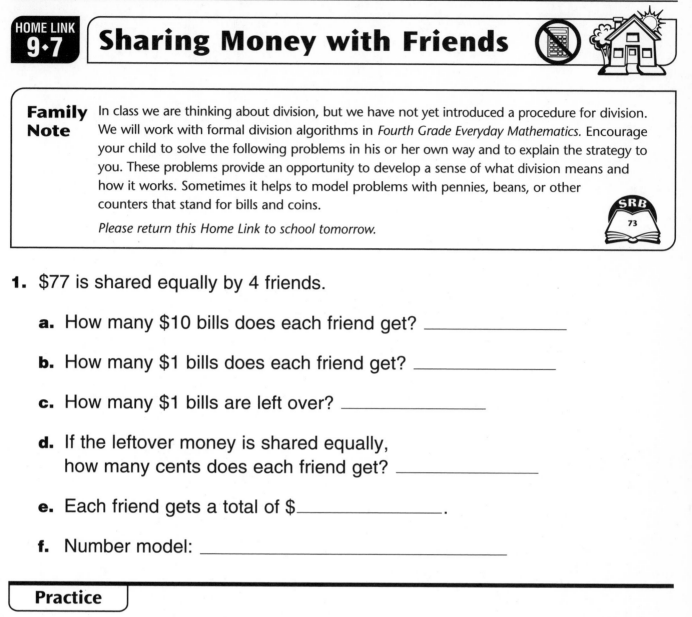

HOME LINK 9·7 **Sharing Money with Friends**

Family Note In class we are thinking about division, but we have not yet introduced a procedure for division. We will work with formal division algorithms in *Fourth Grade Everyday Mathematics.* Encourage your child to solve the following problems in his or her own way and to explain the strategy to you. These problems provide an opportunity to develop a sense of what division means and how it works. Sometimes it helps to model problems with pennies, beans, or other counters that stand for bills and coins.

Please return this Home Link to school tomorrow.

SRB 73

1. $77 is shared equally by 4 friends.

   a. How many $10 bills does each friend get? _____

   b. How many $1 bills does each friend get? _____

   c. How many $1 bills are left over? _____

   d. If the leftover money is shared equally, how many cents does each friend get? _____

   e. Each friend gets a total of $_____.

   f. Number model: _____

Practice

Use the partial-products method to solve these problems. Show your work.

2.　　21
　　× 2

3.　　48
　　× 4

4.　　63
　　× 5

HOME LINK
9·8 | **Equal Shares and Equal Parts**

Family Note As the class continues to investigate division, we are looking at remainders and what they mean. The focus of this assignment is on figuring out what to do with the remainder, NOT on using a division algorithm. Encourage your child to draw pictures, use a calculator, or use counters to solve the problems.

Please return this Home Link to school tomorrow.

Solve the problems below. Remember that you will have to decide what the remainder means in order to answer the questions. You may use your calculator, counters, or pictures to help you solve the problems.

1. There are 31 children in Dante's class. Each table in the classroom seats 4 children. How many tables are needed to seat all of the children?

   _____

2. Emily and Linnea help out on their uncle's chicken farm. One day the hens laid a total of 85 eggs. How many cartons of a dozen eggs could they fill?

   _____

3. Ms. Jerome is buying markers for a scout project. She needs 93 markers. If markers come in packs of 10, how many packs must she buy?

   _____

Practice

Solve each problem using the partial-products algorithm.
Use the back of this Home Link.

4. $29 \times 4 =$ _____   5. $85 \times 5 =$ _____   6. $96 \times 8 =$ _____

215

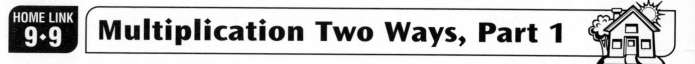

HOME LINK 9·9 Multiplication Two Ways, Part 1

Family Note   Observe as your child solves these problems. See if your child can use more than one method of multiplication, and find out which method your child prefers. Both methods are discussed in the *Student Reference Book* on pages 68–72 and in the Unit 9 Family Letter.

Please return this Home Link to school tomorrow.

SRB 68–72

Use the lattice method and the partial-products algorithm.

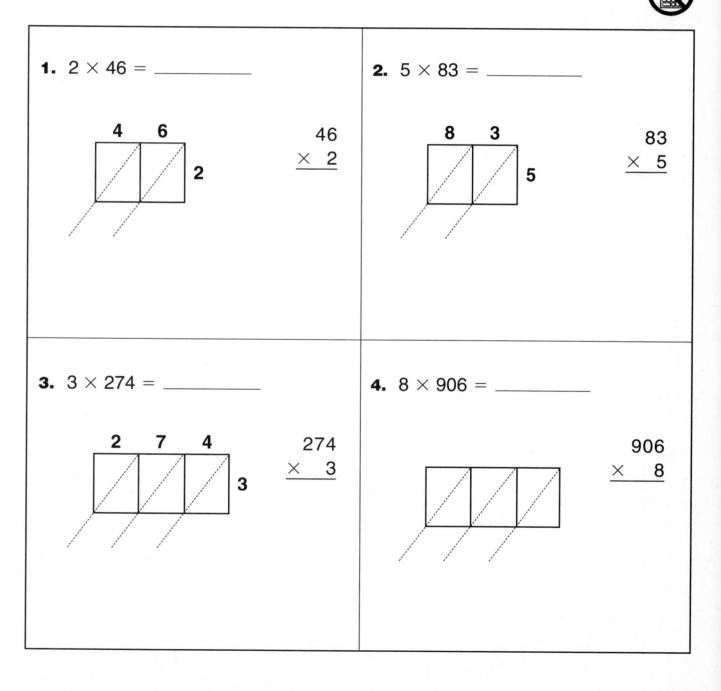

1. 2 × 46 = _____

```
    4   6
  ┌───┬───┐
  │ ╱ │ ╱ │ 2
  └───┴───┘
```
```
   46
 ×  2
```

2. 5 × 83 = _____

```
    8   3
  ┌───┬───┐
  │ ╱ │ ╱ │ 5
  └───┴───┘
```
```
   83
 ×  5
```

3. 3 × 274 = _____

```
    2   7   4
  ┌───┬───┬───┐
  │ ╱ │ ╱ │ ╱ │ 3
  └───┴───┴───┘
```
```
   274
 ×   3
```

4. 8 × 906 = _____

```
  ┌───┬───┬───┐
  │ ╱ │ ╱ │ ╱ │
  └───┴───┴───┘
```
```
   906
 ×   8
```

217

Multiplication Two Ways, Part 2

Family Note The class continues to practice the partial-products algorithm and the lattice method. Encourage your child to try these problems both ways and to compare the answers to be sure they are correct.

Please return this Home Link to school tomorrow.

SRB
68–72

Show someone at home how to use both the lattice method and the partial-products algorithm.

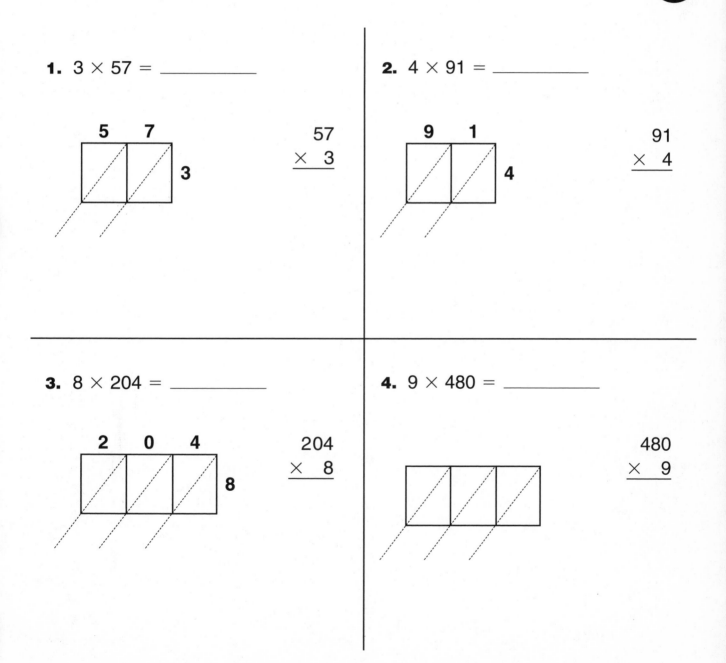

1. 3 × 57 = _____

```
  5   7
┌───┬───┐
│  /│  /│
│ / │ / │ 3
└/──┴/──┘
/    /
```
```
   57
×   3
```

2. 4 × 91 = _____

```
  9   1
┌───┬───┐
│  /│  /│
│ / │ / │ 4
└/──┴/──┘
/    /
```
```
   91
×   4
```

3. 8 × 204 = _____

```
  2   0   4
┌───┬───┬───┐
│  /│  /│  /│
│ / │ / │ / │ 8
└/──┴/──┴/──┘
/    /    /
```
```
  204
×   8
```

4. 9 × 480 = _____

```
  ┌───┬───┬───┐
  │  /│  /│  /│
  │ / │ / │ / │
  └/──┴/──┴/──┘
  /    /    /
```
```
  480
×   9
```

HOME LINK 9·11 — 2-Digit Multiplication: Two Ways

Family Note Your child's class continues to practice the partial-products algorithm and the lattice method, now with 2-digit numbers and 2-digit multiples of 10.

Please return this Home Link to school tomorrow.

SRB 68–72

Use the lattice method and the partial-products algorithm.

1. 20 × 38 = _____

2. 50 × 17 = _____

3. 90 × 62 = _____

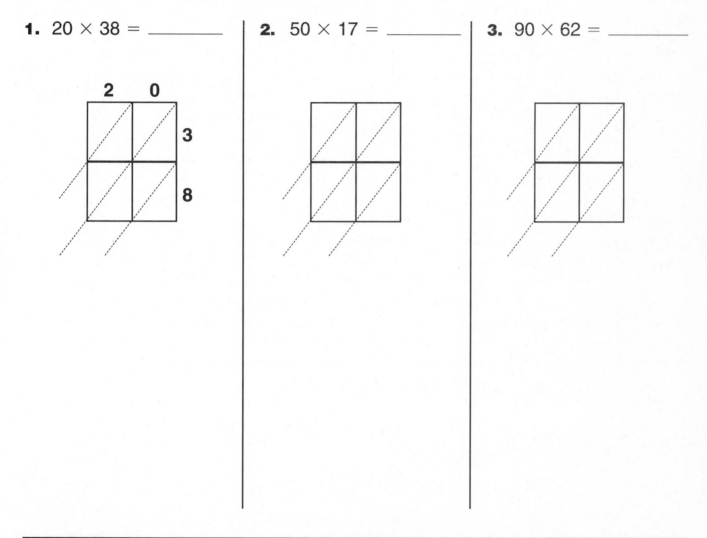

Practice

On the back of this page, use your favorite method to solve these problems.

4. 40 × 28 = _____

5. 60 × 35 = _____

HOME LINK 9·12 | 2 Digits × 2 Digits

Family Note The class continues to practice the partial-products algorithm and the lattice method, now with any 2-digit numbers. Encourage your child to try these problems both ways and to compare the answers to be sure they are correct.

Please return this Home Link to school tomorrow.

SRB 68–72

Use the lattice method and the partial-products algorithm.

1. 21 × 35 = _____

2. 17 × 43 = _____

3. 58 × 62 = _____

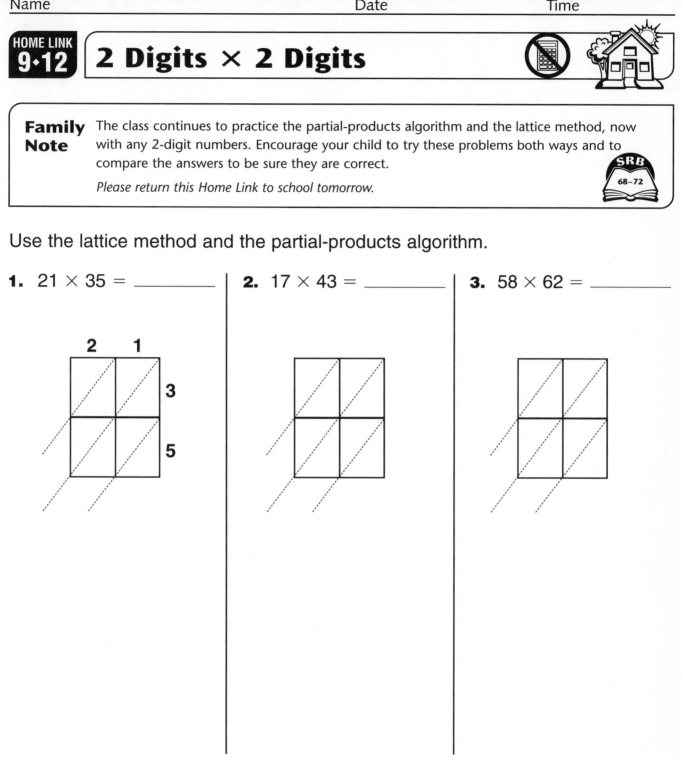

Practice

On the back of this page, use your favorite method to solve these problems.

4. 55 × 49 = _____

5. 91 × 33 = _____

HOME LINK 9·13 Positive and Negative Temperatures

Family Note Encourage your child to use the thermometer pictured here to answer questions about thermometer scales, temperature changes, and temperature comparisons. If you have a real thermometer, try to show your child how the mercury moves up and down.

SRB 170–173

Please return this Home Link to school tomorrow.

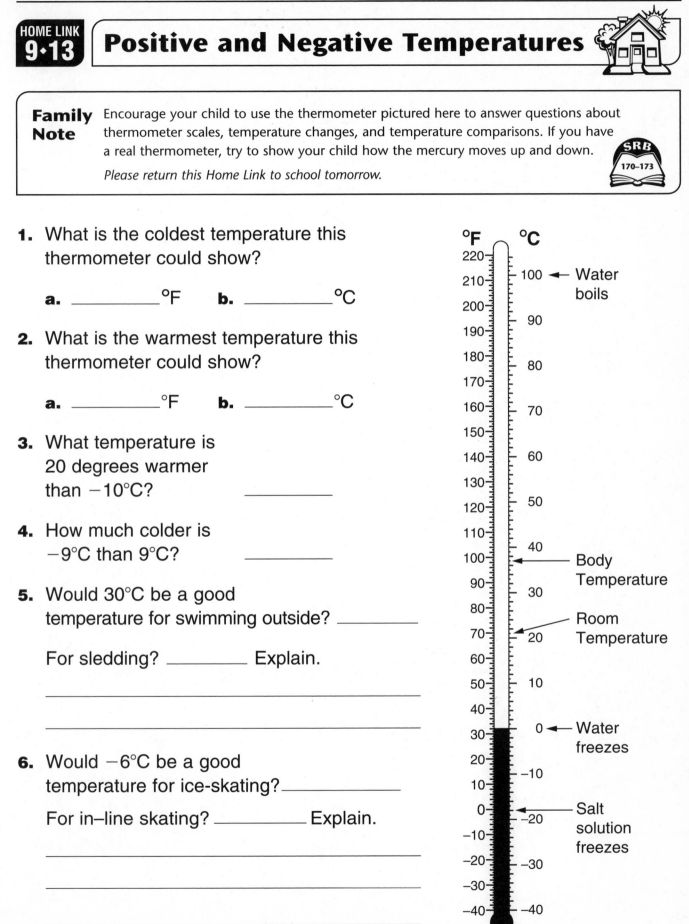

1. What is the coldest temperature this thermometer could show?

   a. _____ °F   b. _____ °C

2. What is the warmest temperature this thermometer could show?

   a. _____ °F   b. _____ °C

3. What temperature is 20 degrees warmer than −10°C? _____

4. How much colder is −9°C than 9°C? _____

5. Would 30°C be a good temperature for swimming outside? _____

   For sledding? _____ Explain.

   _____

   _____

6. Would −6°C be a good temperature for ice-skating? _____

   For in–line skating? _____ Explain.

   _____

   _____

°F   °C

220 —
210 — — 100 ← Water boils
200 —
190 — — 90
180 — — 80
170 —
160 — — 70
150 —
140 — — 60
130 —
120 — — 50
110 —
100 — — 40 ← Body Temperature
90 — — 30
80 —
70 — — 20 ← Room Temperature
60 —
50 — — 10
40 —
30 — — 0 ← Water freezes
20 —
10 — — −10
0 — — −20 ← Salt solution freezes
−10 —
−20 — — −30
−30 —
−40 — — −40

225

Unit 10: Family Letter

Measurement and Data

This unit has three main objectives:

◆ To review and extend previous work with measures of length, weight, and capacity by providing a variety of hands-on activities and applications. These activities will provide children with experience using U.S. customary and metric units of measurement.

◆ To extend previous work with the median and mode of a set of data and to introduce the mean (average) of a set of data.

◆ To introduce two new topics: finding the volume of rectangular prisms and using ordered pairs to locate points on a coordinate grid.

Children will repeat the personal measurements they made earlier in the year so that they may record their own growth. They will display these data in graphs and tables and find typical values for the class by finding the median, mean, and mode of the data.

They will begin to work with volumes of rectangular boxes, which have regular shapes, and will also compare the volumes of several irregular objects and investigate whether there is a relationship between the weight of these objects and their volumes.

Tables of Measures	
Length	1 kilometer = 1,000 meters 1 meter = 100 centimeters 1 centimeter = 10 millimeters 1 mile = 1,760 yards 1 yard = 3 feet 1 foot = 12 inches
Weight	1 kilogram = 1,000 grams 1 gram = 1,000 milligrams 1 ton = 2,000 pounds 1 pound = 16 ounces
Volume & Capacity	1 liter = 1,000 milliliters 1 gallon = 4 quarts 1 quart = 2 pints 1 pint = 2 cups 1 cubic yard = 27 cubic feet 1 cubic foot = 1,728 cubic inches

Please keep this Family Letter for reference as your child works through Unit 10.

Vocabulary

Important terms in Unit 10:

coordinate grid A reference frame for locating points in a plane by means of ordered pairs of numbers. A rectangular coordinate grid is formed by two number lines that intersect at right angles at their zero points.

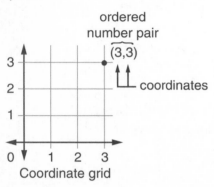

ordered
number pair
(3,3)
coordinates

Coordinate grid

coordinate A number used to locate a point on a number line; a point's distance from an origin.

ordered number pair A pair of numbers used to locate a point on a coordinate grid.

height of a prism The length of the shortest line segment from a base of a prism to the plane containing the opposite face. The height is perpendicular to the base.

volume The amount of space occupied by a 3-dimensional shape.

square centimeter (square cm, cm²) A unit to measure area. For example, a square centimeter is the area of a square with 1-cm long sides.

cubic centimeter (cubic cm, cm³) A metric unit of volume or capacity equal to the volume of a cube with 1cm edges.

weight A measure of how heavy something is; the force of gravity on an object.

capacity (of a scale) The maximum weight a scale can measure. For example, most infant scales have a capacity of about 25 pounds.

capacity (of a container) The amount a container can hold. Capacity is often measured in units such as quarts, gallons, cups, or liters.

frequency table A table in which data are tallied and organized, often as a first step toward making a frequency graph.

Waist–to–floor measurement (inches)	Frequency	
	Tallies	Number
27	//	2
28		0
29	⊬⊬⊬	5
30	⊬⊬⊬ ///	8
31	⊬⊬⊬ //	7
32	////	4
	Total = 26	

mode The value or values that occur most often in a set of data. For example, in the frequency table above, 30 inches is the mode.

mean The sum of a set of numbers divided by the number of numbers in the set. Often called the average value of the set.

Do-Anytime Activities

To work with your child on the concepts taught in this unit and in previous units, try these interesting and rewarding activities:

1. Review equivalent names for measurements. For example: *How many inches in 1 foot? How many pints in 3 quarts? How many centimeters in 1 meter? How many grams in 1 kilogram?*

2. Review multiplication facts. For example: *How much is 6 times 3? 7 × 8? 4 [5s]?*

3. Review division facts. For example: *How many 2s in 12? What number multiplied by 4 equals 12? How much is 18 divided by 2?*

4. Practice multiplication with multiples of 10, 100, and 1,000. For example: *How much are 10 [30s]? How much is 4 × 100? What number times 100 equals 4,000?*

5. Practice division with multiples of 10, 100, and 1,000. For example: *How much is $\frac{1}{10}$ of 300? How many 50s in 5,000? How much is 200 divided by 50?*

Building Skills through Games

In Unit 10, your child will practice mental-math skills by playing the following games:

Memory Addition/Subtraction

Partners agree on a target number. They take turns adding or subtracting any number from 1 to 5 into the memory of their calculators while keeping track of the sums or differences in their heads. Then they press the MRC key to see if the final memory sums match their initial target number.

Multiplication Top-It

Players turn over two cards and call out the product. The player with the higher product keeps all the cards. The player with more cards at the end wins! *You will receive more detailed directions for* Multiplication Top-It *when we begin to play it in class.*

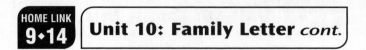

As You Help Your Child with Homework

As your child brings home assignments, you may want to go over the instructions together, clarifying them as necessary. The answers listed below will guide you through this unit's Home Links.

Home Link 10◆1

1. 60; 96

2. 9; 12; 17

3. 33; 6; 12

4. 2; 4; 6

5. $\frac{1}{2}$; $\frac{1}{320}$; $\frac{1}{8}$; $\frac{1}{4}$; $\frac{1}{2}$

6. 90; 152; 117

Home Link 10◆2

1. Boxes B, C, and D

2. Answers vary.

3. Answers vary.

Home Link 10◆3

1. 2,052

2. 3,854

Home Link 10◆5

1. inch

2. gram

3. square yard

4. centimeter

5. inch

6. quart

7. 1 Liter

8. 140

9. 186

10. 864

Home Link 10◆6

4. 3

Home Link 10◆7

1. 60.3

2. 12.8

Home Link 10◆8

1. $20 \times 30 = 600$

$30 \times 20 = 600$

$600 \div 30 = 20$

$600 \div 20 = 30$

2. $40 \times 20 = 800$

$20 \times 40 = 800$

$800 \div 40 = 20$

$800 \div 20 = 40$

3. $100 \times 5 = 500$

$5 \times 100 = 500$

$500 \div 100 = 5$

$500 \div 5 = 100$

4. $600 \times 7 = 4,200$

$7 \times 600 = 4,200$

$4,200 \div 600 = 7$

$4,200 \div 7 = 600$

Home Link 10◆10

1. (3,6) Algeria **2.** (6,3) Tanzania **3.** (5,5) Sudan

4. (4,5) Chad **5.** (5,6) Egypt **6.** (4,6) Libya

HOME LINK 10·1 | Old-Fashioned Equivalencies

Solve the problems yourself. Write your answers on the "slate."

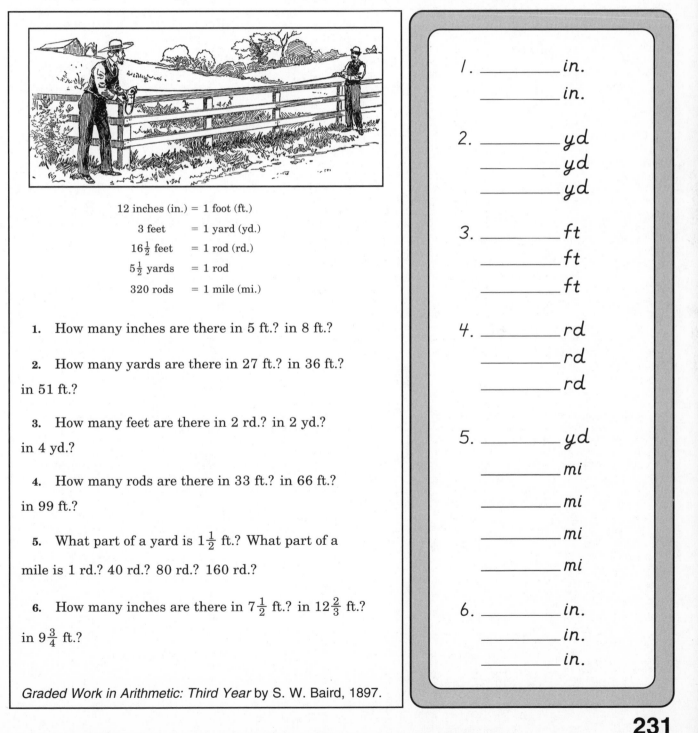

12 inches (in.) = 1 foot (ft.)

3 feet	= 1 yard (yd.)
$16\frac{1}{2}$ feet	= 1 rod (rd.)
$5\frac{1}{2}$ yards	= 1 rod
320 rods	= 1 mile (mi.)

1. How many inches are there in 5 ft.? in 8 ft.?

2. How many yards are there in 27 ft.? in 36 ft.? in 51 ft.?

3. How many feet are there in 2 rd.? in 2 yd.? in 4 yd.?

4. How many rods are there in 33 ft.? in 66 ft.? in 99 ft.?

5. What part of a yard is $1\frac{1}{2}$ ft.? What part of a mile is 1 rd.? 40 rd.? 80 rd.? 160 rd.?

6. How many inches are there in $7\frac{1}{2}$ ft.? in $12\frac{2}{3}$ ft.? in $9\frac{3}{4}$ ft.?

Graded Work in Arithmetic: Third Year by S. W. Baird, 1897.

1. _____ in.
   _____ in.

2. _____ yd
   _____ yd
   _____ yd

3. _____ ft
   _____ ft
   _____ ft

4. _____ rd
   _____ rd
   _____ rd

5. _____ yd
   _____ mi
   _____ mi
   _____ mi
   _____ mi

6. _____ in.
   _____ in.
   _____ in.

HOME LINK
10·2

Exploring the Volume of Boxes

Family Note To explore the concept of volume, our class built open boxes out of patterns like the ones in this Home Link and then filled the boxes with centimeter cubes. Your child should try to calculate the volume of the boxes he or she builds on this Home Link by imagining that it is filled with cubes. Then have your child check the results by pouring a substance from one box to the other, as described below.

SRB
157–159

Please return this Home Link to school tomorrow.

1. Cut out the patterns. Tape or glue each pattern to make an open box. Find boxes that have the same volume.

2. How did you figure out your answer?

_____

_____

_____

_____

_____

_____

_____

_____

3. Check your answer by pouring rice, dried beans, or sand into one of the boxes. Fill the box to the top and level it off with a straightedge like an index card or a ruler. Then pour it into another box. Explain how you know when the boxes have the same volume.

_____

_____

_____

Exploring the Volume of Boxes *cont.*

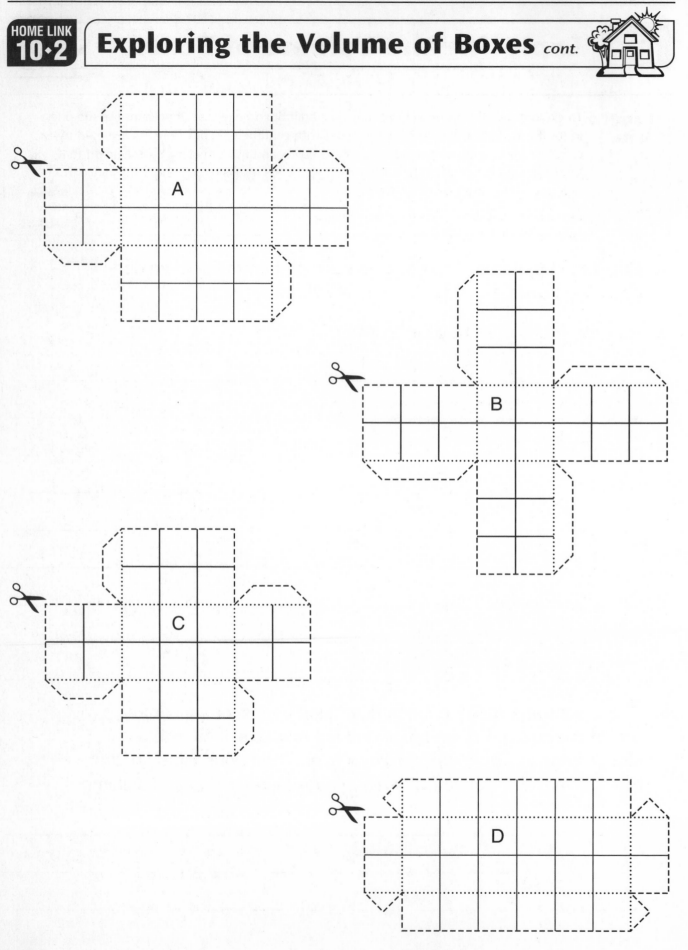

HOME LINK 10·3 The Meaning of Weight

Family Note Today the children discussed weight. They examined different scales, discussed objects that might be weighed with each kind of scale, and read weights on scales.

Please return this Home Link to school tomorrow.

SRB
68–72

Which do you think weighs more: a pound of feathers or a pound of books? Explain your reason.

_____

_____

Practice

Solve each problem using the partial products and lattice algorithms.

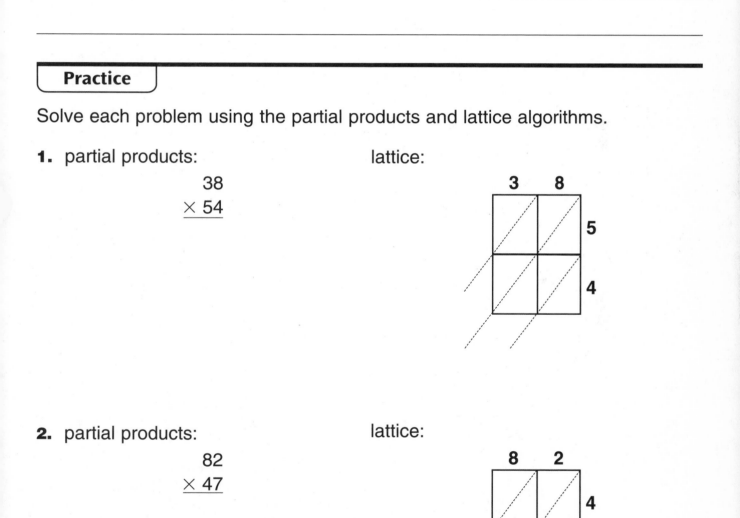

1. partial products:

$$\begin{array}{r} 38 \\ \times\ 54 \\ \hline \end{array}$$

lattice:

```
      3     8
   ┌─────┬─────┐
   │    ╱│    ╱│ 5
   │  ╱  │  ╱  │
   ├─────┼─────┤
   │    ╱│    ╱│ 4
   │  ╱  │  ╱  │
   └─────┴─────┘
```

2. partial products:

$$\begin{array}{r} 82 \\ \times\ 47 \\ \hline \end{array}$$

lattice:

```
      8     2
   ┌─────┬─────┐
   │    ╱│    ╱│ 4
   │  ╱  │  ╱  │
   ├─────┼─────┤
   │    ╱│    ╱│ 7
   │  ╱  │  ╱  │
   └─────┴─────┘
```

HOME LINK 10·4 Collecting Food Container Labels

> **Family Note**
>
> Today our class measured the weight and volume of several objects. We tried to decide whether an object that weighs more than another object always has the greater volume. Ask your child, "Which takes up more space, a pound of popped popcorn or a pound of marbles?"
>
> Help your child practice multiplication facts by playing the game *Multiplication Top-It*. Directions for the game are below.
>
> *Please send the collected food labels to school tomorrow.*
>
> SRB
> 297 298

A. Ask someone at home to help you find food containers showing nutritional information. For example, you might look on canned goods, cereal boxes, bags of cookies, or bottles of cooking oil. Bring the labels or empty containers to school. Be sure they are clean.

B. Play a game of *Multiplication Top-It* with 1 or 2 people at home. *Multiplication Top-It* is similar to the card game *War*.

Directions

1. Remove the face cards from a regular deck of cards. The aces are the 1-cards.

2. Shuffle the cards. Place the deck facedown on a table.

3. Each player turns over two cards and calls out the product of the numbers. The player with the higher product wins the round and takes all the cards.

4. In case of a tie, each player turns over two more cards and calls out the product. The player with the higher product then takes all the cards from both plays.

5. Play ends when not enough cards are left for both players to turn over two cards. The player with more cards wins.

Example Colleen turns over a 6 and a 2. She calls out 12.
Danny turns over a 10 and a 4. He calls out 40.
Danny has the higher product. He takes all 4 cards.

HOME LINK 10·5 Matching Units of Measure

Family Note Today our class explored units of capacity—cups, pints, quarts, gallons, milliliters, and liters. For the list below, your child should choose an appropriate unit for measuring each item. Some of the items refer to capacity, but units of length, weight, area, and volume are also included. Do not expect your child to know all of the units. Remind your child that *square units* refer to area measurement and *cubic units* to volume measurement.

Please return this Home Link to school tomorrow.

SRB
146 154
157 160
162

Fill in the oval to mark the unit best used to measure each object.

	Object	Units		
1.	height of a chair	⬭ mile	⬭ inch	⬭ pound
2.	weight of a penny	⬭ pound	⬭ inch	⬭ gram
3.	area of a football field	⬭ square inch	⬭ square yard	⬭ cubic meter
4.	perimeter of your journal	⬭ kilometer	⬭ gallon	⬭ centimeter
5.	diameter of a dinner plate	⬭ foot	⬭ cubic centimeter	⬭ inch
6.	amount of juice in a carton	⬭ meter	⬭ quart	⬭ square liter

7. About how much water could you drink in 1 day?

⬭ 1 cup   ⬭ 1 milliliter   ⬭ 1 liter   ⬭ 1 gallon

Practice

Solve.

8. 35
     × 4
    ____

9. 62
     × 3
    ____

10. 27
     × 32
     ____

HOME LINK 10·6

Mean, or Average, Number of Fish

Family Note Many of us learned that to find the mean (average) of a set of numbers, we add all the numbers and then divide the total by how many numbers we added. In today's lesson, the class tried a different method of finding the mean. After your child has completed the page, ask him or her to explain how this method works. In the next lesson, we will introduce finding the mean by adding the numbers and dividing to find the answer.

Please return this Home Link to school tomorrow.

The table below lists how many goldfish each child won at the school fun fair.

Name	Number of Goldfish
Reba	3
Bill	1
Lucy	7
Meg	0
Nate	5
Pat	2

1. Put a penny over each shaded square in the bar graph.

2. Move the pennies so that each column has the same number of pennies.

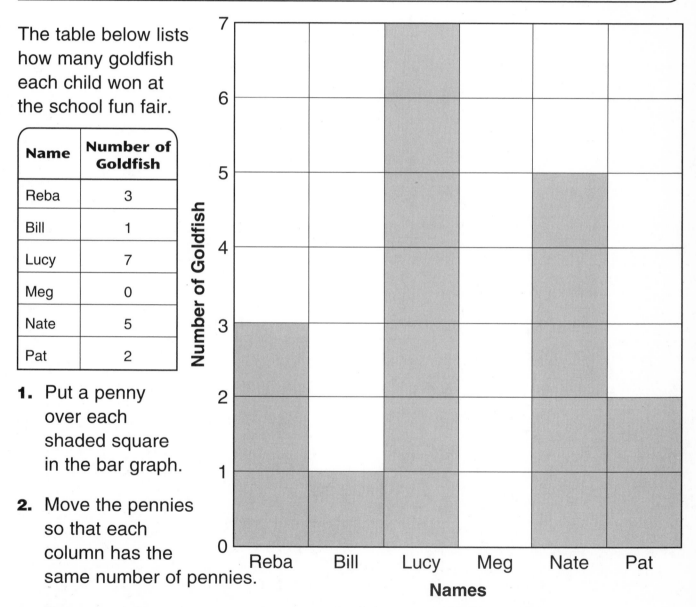

3. Draw a horizontal line across your graph to show the height of the pennies when all of the columns are the same height.

4. The mean (average) number of goldfish won by children at the fun fair is _____.

241

HOME LINK
10·7

Finding the Mean

Family Note The median and mean (average) indicate typical values in a set of data. The median is the middle value when the data numbers are listed in order. The mean (average) is found by the process described below. Your child may use a calculator to solve the problems. (In third grade, we ignore any digits to the right of the tenths place.)

Please return this Home Link to school tomorrow.

SRB
80
83–85

To find the mean (average):

1. Find the sum of the data numbers.

2. Count the data numbers.

3. Use a calculator to divide the sum by the number of data numbers.

4. Drop any digits after tenths.

Example:

Basketball Scores: 80, 85, 76

1. 80 + 85 + 76 = 241

2. There are 3 scores.

3. 241 ÷ 3 = 80.333333...

4. Mean: 80.3

Baseball Home Run Leaders		
1998	Mark McGwire	70
1999	Mark McGwire	65
2000	Sammy Sosa	50
2001	Barry Bonds	73
2002	Alex Rodriguez	57
2003	Jim Thome, Alex Rodriguez	47

1. Mean number of home runs: _____

Baseball Home Run Leaders		
1901	Sam Crawford	16
1902	Socks Seybold	16
1903	Buck Freeman	13
1904	Harry Davis	10
1905	Fred Odwell	9

2. Mean number of home runs: _____

Source: World Almanac, 2004

3. List some data for people in your home—for example, their ages, shoe sizes, or heights. Find the median and mean of the data.

Kind of data _____

Data _____

Median: _____ Mean: _____

HOME LINK 10·8 Fact Triangles

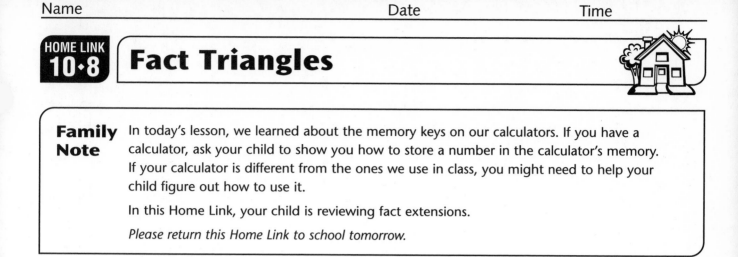

Family Note In today's lesson, we learned about the memory keys on our calculators. If you have a calculator, ask your child to show you how to store a number in the calculator's memory. If your calculator is different from the ones we use in class, you might need to help your child figure out how to use it.

In this Home Link, your child is reviewing fact extensions.

Please return this Home Link to school tomorrow.

Fill in the missing number in each Fact Triangle. Then write the number families for the three numbers in the Fact Triangle.

1.

×, ÷

20   30

_____

_____

_____

2.

800

×, ÷

_____   40

_____

_____

_____

3.

500

×, ÷

5   _____

_____

_____

_____

4.

×, ÷

600   7

_____

_____

_____

HOME LINK
10·9

A Frequency Table

1. Make a frequency table for the number of electrical outlets in at least 8 different rooms.

Number of Electrical Outlets

Room	Frequency	
	Tallies	**Number**

2. What is the *median* (middle) number of outlets? _____

3. What is the *mean* (average) number of outlets? (You may use a calculator to calculate the answer. Drop any digits to the right of the tenths place.) _____

4. What is the *mode* of the data in the table? (*Reminder:* The mode is the number that occurs most often in a set of data.) _____

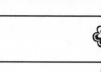

HOME LINK 10·10 Locating Points on a Map

Family Note In an ordered pair, such as (3,6), the first number indicates how far the point is to the right (or left) of 0. The second number indicates how far it is above (or below) 0.

Please return this Home Link to school tomorrow.

SRB
180 181

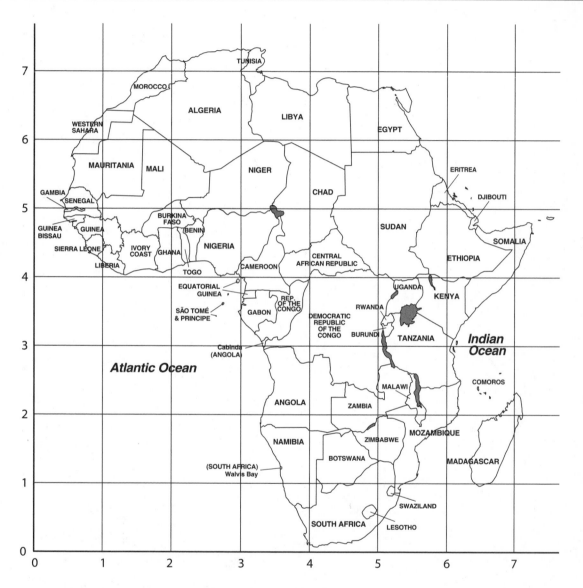

Here is a map of Africa. Write the name of the country in which each point is located.

1. (3,6) _____

2. (6,3) _____

3. (5,5) _____

4. (4,5) _____

5. (5,6) _____

6. (4,6) _____

249

Unit 11: Family Letter

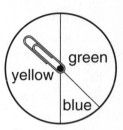

Probability; Year-Long Projects Revisited

In this year's final unit, children will have the opportunity to bring closure to the yearlong data-collection projects about lengths of days and temperature changes. They will look at patterns in data and draw conclusions.

Unit 11 also contains informal spinner activities relating to chance and probability.

Some of these activities call for children to compare the likelihood of several possible outcomes of an event: why one thing is more likely to happen than another. For example, children will make predictions about where a paper clip on a spinner is more likely to land when the spinner is divided into unequal parts.

Other activities ask children to estimate the chance that something will happen. For example, children design a spinner so that a paper clip is twice as likely to land on one color as another.

Please keep this Family Letter for reference as your child works through Unit 11.

Vocabulary

Important terms in Unit 11:

equally likely outcomes Outcomes of a chance experiment or situation that have the same probability of happening. For example, any number 1–6 landing up are the equally likely outcomes of rolling a die.

winter solstice The shortest day of the year, when the sun is farthest south of the Earth's equator. The number of hours of daylight depends on your latitude. In the Northern Hemisphere, the winter solstice occurs on or about December 21.

summer solstice The longest day of the year, when the sun is farthest north of the Earth's equator. The number of hours of daylight depends on your latitude. In the Northern Hemisphere, the summer solstice occurs on or about June 21.

Do-Anytime Activities

To work on the concepts taught in this unit and in previous units, try these interesting and rewarding activities:

1. When you are in the car or walking with your child, search for geometric figures. Identify them by name if possible and talk about their characteristics. For example, a stop sign is an octagon, which has 8 sides and 8 angles. Many skyscrapers are rectangular prisms; their faces are rectangles.

2. Draw name-collection boxes for various numbers and together with your child write five to ten equivalent names in each box. Include name-collection boxes for fractions and decimals. For example, a $\frac{1}{2}$ name-collection box might include $\frac{2}{4}$, $\frac{10}{20}$, 0.5, 0.50, and $\frac{500}{1,000}$ because these are also names for $\frac{1}{2}$. Then create name-collection boxes that include equivalent measures. For example, a 1 ft name-collection box might contain 12 in., $\frac{1}{3}$ yd, $\frac{1}{5,280}$ mile, $\frac{12}{36}$ yd, and so on.

3. Make predictions about the likelihood of pulling an item of one color out of a bag filled with the same items of different colors. Then check your predictions. For example, place 2 red blocks and 4 blue blocks in a bag. There are 4 out of 6 chances to pull a blue block.

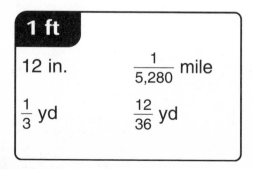

1 ft	
12 in.	$\frac{1}{5,280}$ mile
$\frac{1}{3}$ yd	$\frac{12}{36}$ yd

Building Skills through Games

In Unit 11, your child will practice skills related to chance and probability by playing the following games. For detailed instructions, see the *Student Reference Book*.

Block Drawing Game

Without letting the other players see the blocks, a Director puts five blocks in a paper bag and tells the players how many blocks are in the bag. A player takes a block out of the bag. The Director records the color of the block for all players to see. The player replaces the block. At any time, a player may say *Stop!* and guess how many blocks of each color are in the bag. If a player guesses incorrectly, that player is out of the game. The first player to guess correctly wins the game.

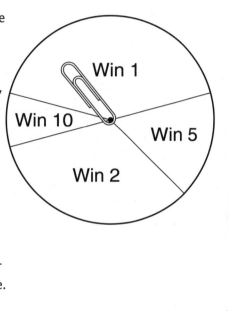

Spinning to Win

Each player claims one section of the spinner. Players take turns spinning the spinner. If the spinner lands on a player's number, the player takes that number of pennies. The player with the most pennies after 12 spins wins the game.

As You Help Your Child with Homework

As your child brings home assignments, you may want to go over the instructions together, clarifying them as necessary. The answers listed below will guide you through this unit's Home Links.

Home Link 11◆2

Numbers	Add	Subtract	Multiply	Divide
30 and 7	37	23	210	4R2
50 and 5	55	45	250	10
40 and 6	46	34	240	6 R4
150 and 3	153	147	450	50
3,000 and 50	3,050	2,950	150,000	60
12,000 and 60	12,060	11,940	720,000	200

Home Link 11◆5

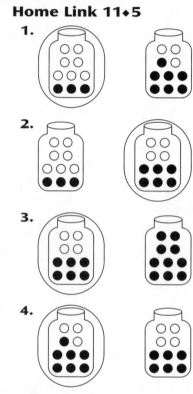

1.
2.
3.
4.

HOME LINK 11·1

A Survey

Family Note

Have your child survey 10 people—family members, neighbors, and out-of-school friends—to find out how many are right-handed and how many are left-handed. Do not count people who say they are ambidextrous (able to use both hands with equal ease). Take a few days to help your child complete the survey. The results will be used in Lesson 11-5.

Please return this Home Link to school.

1. Ask 10 people whether they are right-handed or left-handed. Do not ask people at your school. Do not count people who say they are neither right-handed nor left-handed. (People who can use both hands with equal ease are called *ambidextrous*.)

2. On the chart below, make a tally mark for each person. Be sure that you have exactly 10 marks.

	Tallies
Right-handed	
Left-handed	

3. When you have finished your survey, record the results at the bottom of the page. Bring the results to school.

- ✂

Name _____

Survey Results

Number of right-handed people: _____

Number of left-handed people: _____

Total: 10

255

Computation Round-Up

Family Note   Please observe as your child adds, subtracts, multiplies, and divides pairs of whole numbers. Encourage your child to use and explain his or her favorite strategies.

Please return this Home Link to school.

For each of the number pairs below, use mental arithmetic or other strategies to perform the operations indicated in each column in the table. Show any work on the back of this page. Explain your favorite strategies to someone at home.

Numbers	Add	Subtract	Multiply	Divide
30 and 7	30 + 7 = 37	30 − 7 = 23	30 × 7 = 210	30 ÷ 7 → 4 R2
50 and 5				
40 and 6				
150 and 3				
3,000 and 50				
12,000 and 60				

257

HOME LINK 11·3   A Fair Game?

Family Note   To explore probability, play the game *Fingers* with your child. After 20 games, have your child decide if the game is fair and explain why or why not. (A game is fair if all players have an equal chance of winning or losing.)

Please return this Home Link to school tomorrow.

Play *Fingers* at least 20 times. Keep a tally of wins and losses in the table below.

Rules for *Fingers*

This is a game for 2 players. One player tries to guess the number of fingers the other player will throw (display).

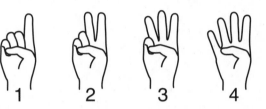

1   2   3   4

You, the *Everyday Mathematics* student, can throw 1, 2, 3, or 4 fingers. The other player can throw only 1 or 2 fingers.

Players face each other. Each one puts a closed fist on his or her chest.

One player counts, "One, two, three." On "three," each player throws some number of fingers.

At the same time, both players call out what they think will be the total number of fingers thrown by both players.

◆ The player who calls out the correct total wins.

◆ If *neither* player calls out the correct total, no one wins.

◆ If *both* players call out the correct total, no one wins.

Tallies for Wins	Tallies for Losses

1. Is this game fair? (Fair means each player has the same chance of winning.) _____

2. On the back of this page explain your answer.

Adaptation of rules for *Mora* in *Family Fun and Games,* The Diagram Group, Sterling Publishing, 1992, p. 365

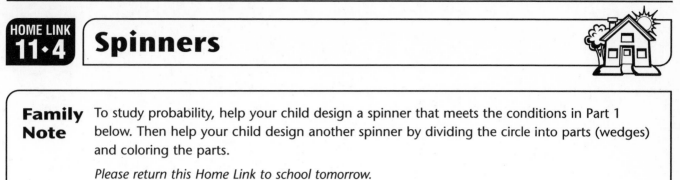

HOME LINK 11·4 Spinners

Family Note To study probability, help your child design a spinner that meets the conditions in Part 1 below. Then help your child design another spinner by dividing the circle into parts (wedges) and coloring the parts.

Please return this Home Link to school tomorrow.

Work with someone at home to make two spinners.

1. Use blue, red, yellow, and green crayons or coloring pencils on the first spinner. Color the spinner so that all of the following are true:

   When spun around a pencil point in the center of the circle, a paper clip

   ◆ is very likely to land on red.

   ◆ has the same chance of landing on yellow as on green.

   ◆ may land on blue but is very unlikely to land on blue.

2. Design and color your own spinner. Then tell how likely or unlikely it is that the paper clip will land on each of the colors you used.

   _____

   _____

   _____

   _____

   _____

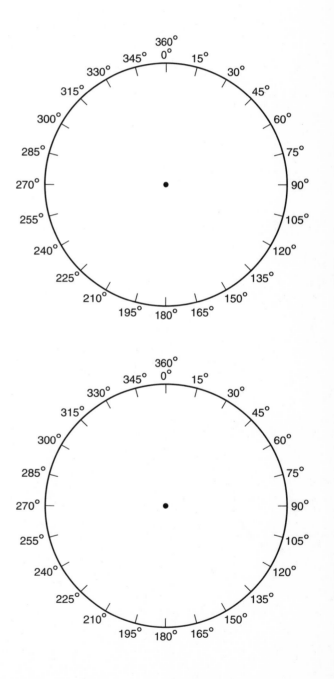

HOME LINK 11·5 More Random-Draw Problems

Family Note This Home Link focuses on predicting the contents of a jar by drawing out marbles. Don't expect your child to be an expert. Explorations with probability will continue through sixth grade. This is a first exposure.

Please return this Home Link to school tomorrow.

In each problem there are 10 marbles in a jar. The marbles are either black or white. A marble is drawn at random (without looking) from the jar. The type of marble drawn is tallied. Then the marble is returned to the jar.

◆ Read the description of the random draws in each problem.

◆ Circle the picture of the jar that best matches the description.

Example: From 100 random draws, you get:

| a black marble | ● | 81 times. |
| a white marble | ○ | 19 times. |

1. From 100 random draws, you get:

| a black marble | ● | 34 times. |
| a white marble | ○ | 66 times. |

2. From 100 random draws, you get:

| a black marble | ● | 57 times. |
| a white marble | ○ | 43 times. |

Try This

3. From 50 random draws, you get:

| a black marble | ● | 28 times. |
| a white marble | ○ | 22 times. |

4. From 50 random draws, you get:

| a black marble | ● | 35 times. |
| a white marble | ○ | 15 times. |

263

Family Letter

Congratulations!

By completing *Third Grade Everyday Mathematics,* your child has accomplished a great deal. Thank you for all of your support!

This Family Letter is here for you to use as a resource throughout your child's summer vacation. It includes an extended list of Do-Anytime Activities, directions for games that can be played at home, a list of mathematics-related books to check out over vacation, and a sneak preview of what your child will be learning in *Fourth Grade Everyday Mathematics.* Enjoy your vacation!

Do-Anytime Activities

Mathematics means more when it is rooted in real-life situations. To help your child review many of the concepts he or she has learned in third grade, we suggest the following activities for you and your child to do together over vacation. These activities will help your child build on the skills he or she has learned this year and help prepare him or her for *Fourth Grade Everyday Mathematics.*

1. If you receive a daily newspaper, continue with the length-of-day project by recording the time of sunrise and sunset once a week. Draw conclusions about the length of a day during vacation months.

2. Over a period of time, have your child record the daily temperatures in the morning and in the evening. Keep track of the findings in chart or graph form. Ask questions about the data—for example, to find the differences in temperatures from morning to evening or from one day to the next.

3. As you are driving in the car or going on walks, search for geometric figures and identify them by name along with some of their characteristics. For example: A stop sign is an octagon, which has eight sides and eight angles; an orange construction cone is a cone, which has one flat surface that is shaped like a circle, a curved surface, and an apex; a brick is a rectangular prism in which all faces are rectangles.

4. Continue to practice addition, subtraction, multiplication, and division facts. Using short drill sessions with Fact Triangles, fact families, and games helps your child build on previous knowledge.

5. Provide multidigit addition and subtraction problems for your child to solve; encourage your child to write number stories to go along with the number models.

Building Skills through Games

The following section lists rules for games that can be played at home. The number cards used in some games can be made from 3" by 5" index cards.

Division Arrays

Materials
- ☐ number cards 6–18 (3 of each)
- ☐ 18 counters, such as pennies
- ☐ 1 regular die
- ☐ scratch paper for each player

Players   2 to 4

Directions

Shuffle the cards and place the deck facedown on the playing surface.

At each turn, a player draws a card and takes the number of counters shown on the card. Next, the player rolls the die. The number on the die specifies the number of equal rows the player must have in the array he or she makes using the counters.

The player's score is the number of counters in each row. If there are no leftover counters, the player's score is double the number of counters in each row.

Players take turns. They keep track of their scores on scratch paper. The player with the highest total at the end of five rounds wins.

Three Addends

Materials
- ☐ paper and pencil (for each player)
- ☐ number cards 1–20 (1 of each)

Players   2

Directions

Shuffle the cards and place the deck facedown on the playing surface.

In turn, players draw three cards from the top of the deck. Both players write addition models using the three numbers on their sheets of paper. (The numbers can be written in whatever order they find easiest for solving the problem.) Players solve the problem and then compare answers.

Option: For a harder version, players take turns drawing four cards from the top of the deck. Players thus solve problems with four addends.

Baseball Multiplication

Materials
- ☐ 2 regular dice
- ☐ 4 pennies
- ☐ score sheet (see below)
- ☐ calculator

Players 2

Directions

Draw a diamond and label *home plate, first base, second base,* and *third base.* Make a score sheet that looks like the one below.

SCORE SHEET

Innings	1	2	3	4	5	6	Total
Player 1 outs							
Runs							
Player 2 outs							
Runs							

1. Take turns being the pitcher and the batter.

2. At the start of the inning, the batter puts a penny on home plate.

3. The pitcher rolls the dice. The batter multiplies the two numbers that come up and tells the answer. The pitcher checks the answer with a calculator.

4. If it is correct, the batter looks up the product in the Hitting Table. The batter either makes an out or moves a penny along the diamond for a single, double, triple, or home run.

   An incorrect solution is a strike, and another pitch (dice roll) is thrown. Three strikes make an out.

> **HITTING TABLE**
>
> 36 = Home Run
> 26–35 = Triple
> 16–25 = Double
> 6–15 = Single
> 5 or less = Out

5. A run is scored each time a penny crosses home plate.

6. A player remains the batter for 3 outs. Then players switch roles. The inning is over when both players have made 3 outs.

7. After making the third out, a batter records the number of runs scored in that inning on the score sheet.

8. The player who has more runs at the end of six innings wins the game.

267

Vacation Reading with a Mathematical Twist

Books can contribute to children's learning by presenting mathematics in a combination of real-world and imaginary contexts. The titles listed below were recommended by teachers who use *Everyday Mathematics* in their classrooms. They are organized by mathematical topic. Visit your local library and check out these mathematics-related books with your child.

Geometry

A Cloak for the Dreamer by Aileen Friedman

Fractals, Googols, and Other Mathematical Tales by Theoni Pappas

Sir Cumference and the First Round Table: A Math Adventure by Wayne Geehan

Measurement

How Tall, How short, How Far Away by David Adler

Math Curse by Jon Scieszka

The Story of Money by Betsy Maestro

If You Made a Million by David Schwartz

Measuring on Penny by Loren Leedy

Numeration

Fraction Fun by David Adler

How Much Is a Million? by David Schwartz

Operations

The Grapes of Math by Gregory Tang

The King's Chessboard by David Birch

The I Hate Mathematics! Book by Marilyn Burns

A Remainder of One by Elinor J. Pinczes

Anno's Mysterious Multiplying Jar by Masqichiro Anno

Patterns, Functions, and Algebra

Eight Hands Round: A Patchwork Alphabet by Ann Whitford Paul

A Million Fish…More or Less by Patricia C. McKissack

Reference Frames

Pigs in a Blanket by Amy Axelrod

Three Days on a River in a Red Canoe by Vera B. Williams

Looking Ahead: *Fourth Grade Everyday Mathematics*

Next year, your child will …

- ◆ go on a World Tour.
- ◆ continue to practice addition and subtraction skills.
- ◆ develop multiplication and division skills.
- ◆ investigate methods for solving problems using mathematics in everyday situations.
- ◆ work with number lines, coordinates, times, latitude and longitude, and dates.
- ◆ collect, organize, and interpret numerical data.
- ◆ continue to explore 3-dimensional objects and their properties, uses, and relationships.
- ◆ read, write, and use whole numbers, fractions, decimals, percents, and negative numbers.
- ◆ explore scientific notation.

Again, thank you for all of your support this year. Have fun continuing your child's mathematics experiences throughout the vacation!